78

Fortschritte der Chemie organischer Naturstoffe

Progress in the Chemistry of Organic Natural Products

Founded by
L. Zechmeister

Edited by
W. Herz, H. Falk,
G. W. Kirby, R. E. Moore,
and Ch. Tamm

Authors:
A. Akhila, G. Adam, K. Rani,
J. Schmidt, and B. Schneider

Springer-Verlag Wien GmbH

Prof. W. HERZ, Department of Chemistry,
The Florida State University, Tallahassee, Florida, U.S.A.

Prof. Dr. H. FALK, Institut für Chemie,
Johannes-Kepler-Universität, Linz, Austria

Prof. G. W. KIRBY, Chemistry Department,
The University of Glasgow, Glasgow, Scotland

Prof. R. E. MOORE, Department of Chemistry,
University of Hawaii at Manoa, Honolulu, Hawaii, U.S.A.

Prof. Dr. CH. TAMM, Institut für Organische Chemie der Universität Basel,
Basel, Switzerland

Originally published by Springer-Verlag Wien New York in 1999
Softcover reprint of the hardcover 1st edition 1999
Library of Congress Catalog Card Number AC 39-1015

Typesetting: Thomson Press (India) Ltd., New Delhi

Graphic design: Ecke Bonk
Printed on acid-free and chlorine-free bleached paper

SPIN: 10725830

With 2 Figures

ISSN 0071-7886
ISBN 978-3-7091-7313-8 **ISBN 978-3-7091-6394-8 (eBook)**
DOI 10.1007/978-3-7091-6394-8

Contents

List of Contributors VIII

Brassinosteroids
By G. ADAM, J. SCHMIDT, and B. SCHNEIDER 1

1. Introduction 2
2. Natural Occurrence and Distribution 2
3. Structures 12
4. Isolation and Purification 13
5. Analysis of Brassinosteroids 13
   5.1. GC-MS Analysis 13
   5.2. HPLC Analysis 21
   5.3. NMR Spectroscopy 21
6. Synthesis 21
7. Biosynthesis 25
   7.1. Biosynthesis in *Catharanthus roseus* 25
   7.2. Biosynthesis in Other Species 27
8. Metabolism of Brassinosteroids 29
   8.1. Metabolism in Plants 29
   8.2. Metabolism in Cell Cultures of *Lycopersicon esculentum* 30
   8.3. Metabolism in Cell Cultures of *Ornithopus sativus* 31
9. Physiological Action 33
10. Molecular Mode of Action 34
11. Conclusions 36

References 36

Chemistry of the Neem Tree (*Azadirachta indica* A. Juss.)
By A. AKHILA and K. RANI 47

1. Introduction 48
2. Chemistry of Limonoids 50
   2.1. Protolimonoids 50

2.2. Apo-Protolimonoids 56
2.3. Apo-Protolimonoids Derived from Loss of 4C-Atoms from the Side Chain which Possess a Hemiacetal Group 59
2.4. Limonoids with Intact Four Rings and a γ-Hydroxybutenolide Side Chain 59
2.5. Azadirone and its Natural Analogues 62
2.6. Homoazadirone Group 69
2.7. Gedunin Group 70
2.8. Vilasinin Group 73

C-Seco Meliacins

2.9. Nimbin Group 81
2.10. Nimbolide Group 85
2.11. Nimbinene Group 87
2.12. Nimbolinin Group 90
2.13. Salannin Group 93
2.14. Azadirachtol Group 97
2.15. Meliacarpin Group 100
2.16. Meliacarpinin and Azadirachtinin Group 100
2.17. Azadirachtin Group 103
2.18. Azadirachtin 109
2.18.1. Biological Activity 109
2.18.2. Structure-Activity Relationships 110
2.18.3. Structure Determination 111
2.18.4. Chemical Reactions 112
2.18.4.1. Reaction of –OH Group 116
2.18.4.1.1. Acetylation 116
2.18.4.1.2. Silylation 116
2.18.4.1.3. Methylation 116
2.18.4.2. Hydrogenation 116
2.18.4.3. Reactions of the Enol Ether Functions 118
2.18.4.4. Saponification Reactions 118
2.18.4.5. Functional Group Chemistry of Azadirachtol 120
2.18.4.6. Oxidation Reactions 120
2.18.4.7. Functional Group Chemistry of 7-Keto Azadirachtins 121
2.18.4.8. Retro-Aldol Reaction 121
2.18.4.9. Skeletal Rearrangements 124
2.18.5. Synthesis 124
2.18.5.1. Synthesis of Dihydrofuranacetal Fragment 'A' 125
2.18.5.1.1. Preparation of Prototype Coupling Fragment 125
2.18.5.2. Decalin 'B' Synthesis 125
2.18.5.3. Coupling of 'A' and 'B' Fragments 126

3. Other Compounds 127
3.1. Diterpenoids 127
3.2. Steroids and Other Triterpenoids 127
3.3. Phenolic Compounds 127
3.3.1. Flavonoids 127

3.3.2. Flavonoglycosides ... 129
3.3.3 Coumarins ... 129
3.3.4. Dihydrochalcone ... 130
3.3.5. Tannins ... 130
3.4. Carbohydrates and Proteins ... 130
3.5. Sulphur Compounds ... 130
3.6. Hydrocarbons, Acids and Esters ... 130

References ... 132

Author Index ... 151

Subject Index ... 159

List of Contributors

ADAM, Prof. Dr. G., Department of Natural Products Chemistry, Institute of Plant Biochemistry, D-06120 Halle (Saale), Germany

AKHILA, Dr. A., Central Institute of Medicinal and Aromatic Plants, P.O. CIMAP, Lucknow 226 015, India

RANI, Dr. K., Central Institute of Medicinal and Aromatic Plants, P.O. CIMAP, Lucknow 226 015, India

SCHMIDT, Dr. J., Department of Natural Products Chemistry, Institute of Plant Biochemistry, D-06120 Halle (Saale), Germany

SCHNEIDER, Dr. B., Research Group Biosynthesis/NMR, Max Planck Institute for Chemical Ecology, D-07745 Jena, Germany

Brassinosteroids

G. ADAM[1], J. SCHMIDT[1], and B. SCHNEIDER[2]
[1] Institute of Plant Biochemistry, Halle/S., Germany
[2] Max-Planck-Institute for Chemical Ecology, Jena, Germany

Contents

1. Introduction 2

2. Natural Occurrence and Distribution 2

3. Structures 12

4. Isolation and Purification 13

5. Analysis of Brassinosteroids 13
   5.1. GC-MS Analysis 13
   5.2. HPLC Analysis 21
   5.3. NMR Spectroscopy 21

6. Synthesis 21

7. Biosynthesis 25
   7.1. Biosynthesis in *Catharanthus roseus* 25
   7.2. Biosynthesis in Other Species 27

8. Metabolism of Brassinosteroids 29
   8.1. Metabolism in Plants 29
   8.2. Metabolism in Cell Cultures of *Lycopersicon esculentum* 30
   8.3. Metabolism in Cell Cultures of *Ornithopus sativus* 31

9. Physiological Action 33

10. Molecular Mode of Action 34

11. Conclusions 36

References 36

1. Introduction

It was in 1979 when GROVE *et al.* isolated from pollen of rape (*Brassica napus*) a highly active plant growth promoter, named it brassinolide and elucidated its structure as (22*R*,23*R*,24*S*)-2α,3α,22,23-tetrahydroxy-24-methyl-B-homo-6a-oxa-5α-cholestan-6-one (**1**) by spectroscopic methods including X-ray analysis (*1*). The original structural features of this compound and its unique high biological activity at very low concentrations stimulated intense research activities in many laboratories. Such efforts were directed towards the search for similar compounds in the plant kingdom, their chemical synthesis, biochemistry and biological mode of action leading up to their practical application in agriculture and horticulture. As a result of this interdisciplinary and rapidly processing research, brassinosteroids can nowadays be regarded as a new class of plant hormones with ubiquitous occurrence in the plant kingdom. Especially, recent molecular biological studies demonstrated their essential role for normal plant growth and development. A series of reviews have been published (*2–11*). Whereas the first book on brassinosteroid research covers developments up to 1990 (*12*), two up-to-date publications about this topic have appeared very recently (*13*, *14*). The present article covers the literature up to December 1998 with special consideration of phytochemical, analytical and biochemical aspects.

2. Natural Occurrence and Distribution

Till now more than 40 members of the brassinosteroid family including conjugates could be structurally identified (Fig. 1, *1*, *15–89*). They have been found in a large variety of higher plants and could be detected in more than 50 species (*3*, *5–10*, *12*, *13*, *20*). Most brassinosteroids described so far have been isolated from plants belonging to the angiosperms and within this subdivision mainly in the class of the dicots, but have also been found in some gymnosperms (Fig. 2).

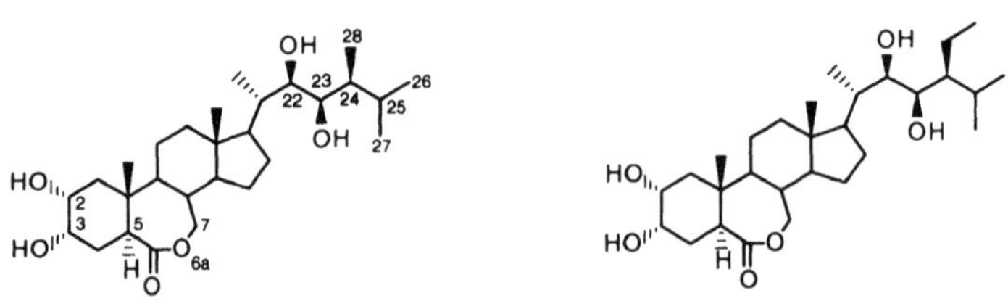

Brassinolide (1) **28-Homobrassinolide (2)**

Fig. 1. Structures of brassinosteroids from plants

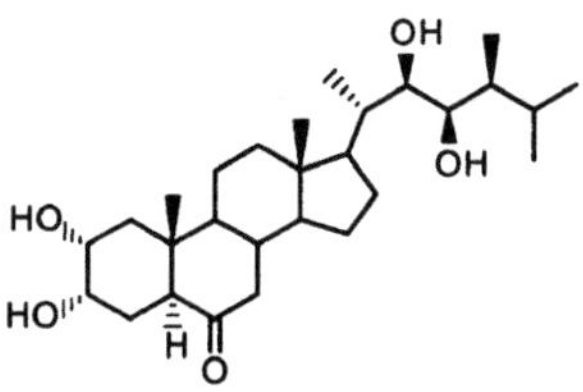

Castasterone (3)

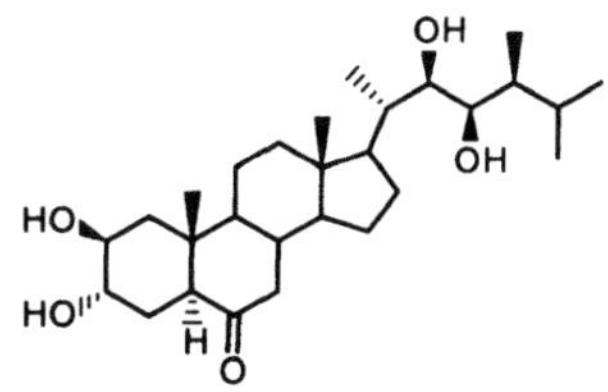

2-Epicastasterone (4)

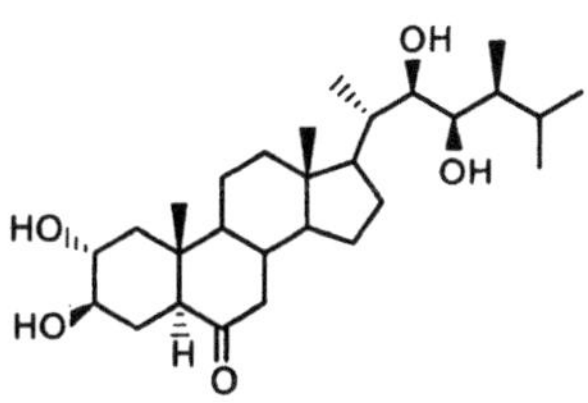

3-Epicastasterone (5)

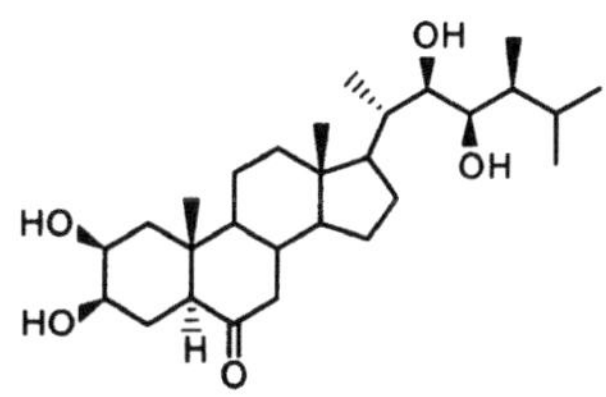

2,3-Diepicastasterone (6)

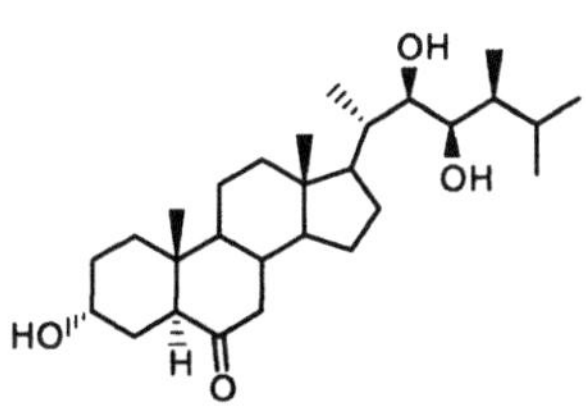

Typhasterol (7)

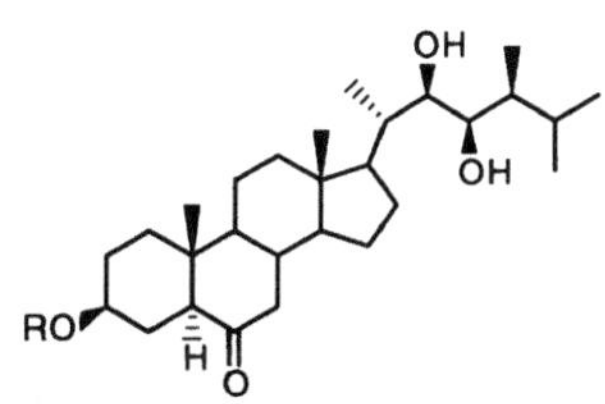

Teasterone (8) : R = H
Teasterone-3-laurate (9) : R = $CO(CH_2)_{10}CH_3$
Teasterone-3-myristate (10) : R = $CO(CH_2)_{12}CH_3$

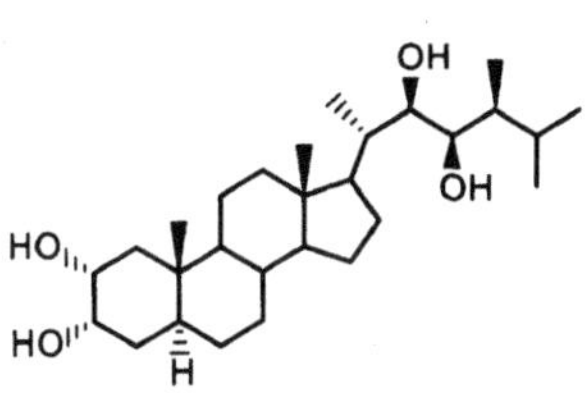

6-Deoxocastasterone (11)

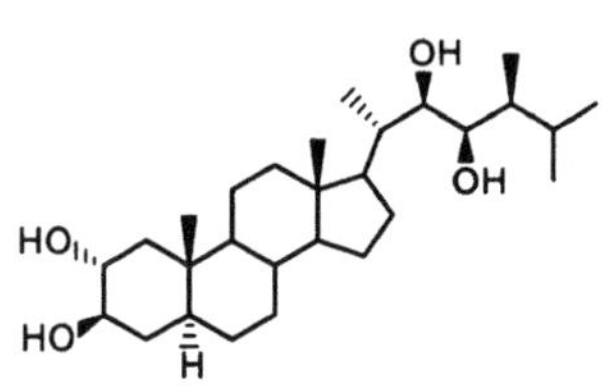

3-Epi-6-deoxocastasterone (12)

Fig. 1 *(continued)*

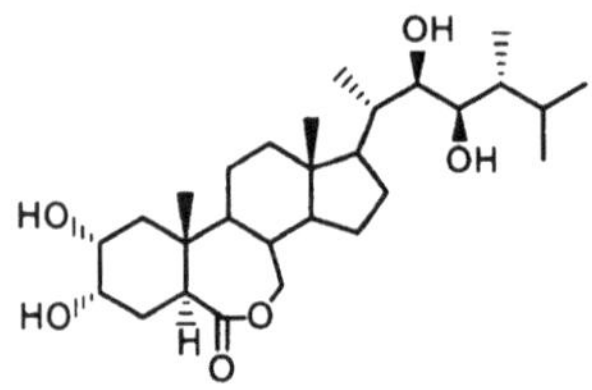

24-Epibrassinolide (13)

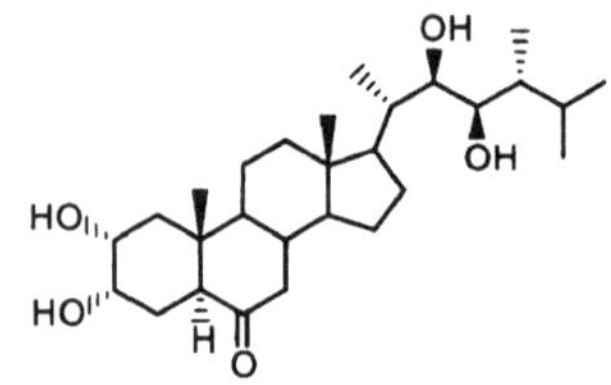

24-Epicastasterone (14)

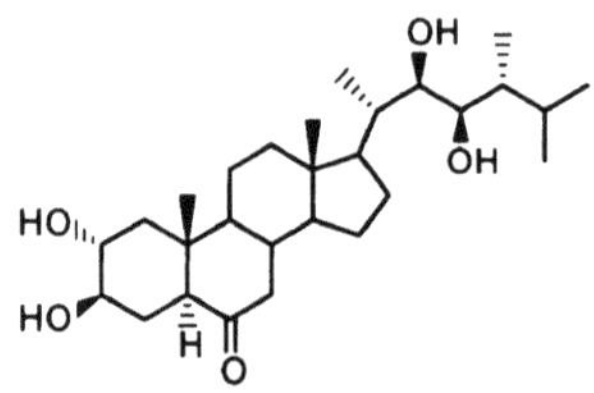

3,24-Diepicastasterone (15)

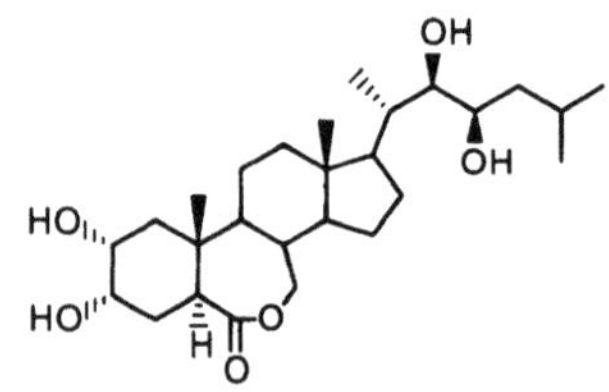

28-Norbrassinolide (16)

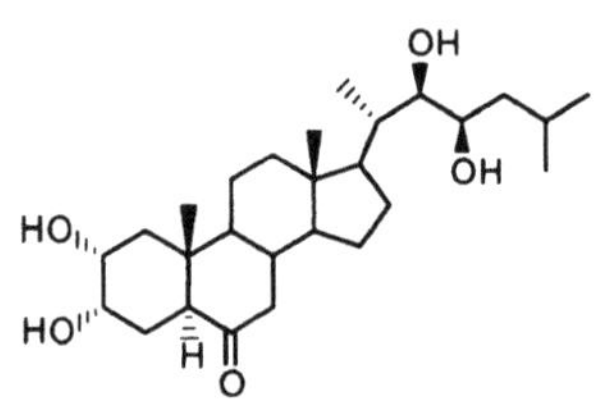

28-Norcastasterone (17)

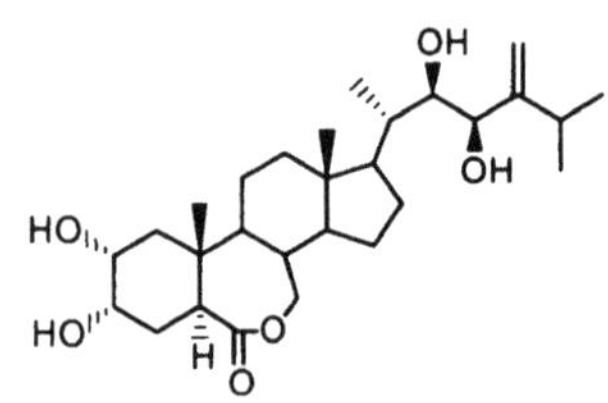

Dolicholide (18)

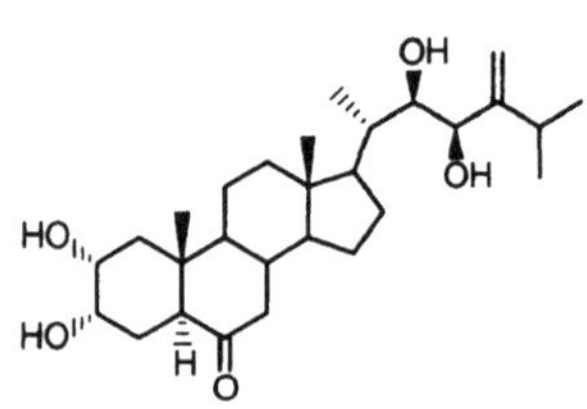

Dolichosterone (19)

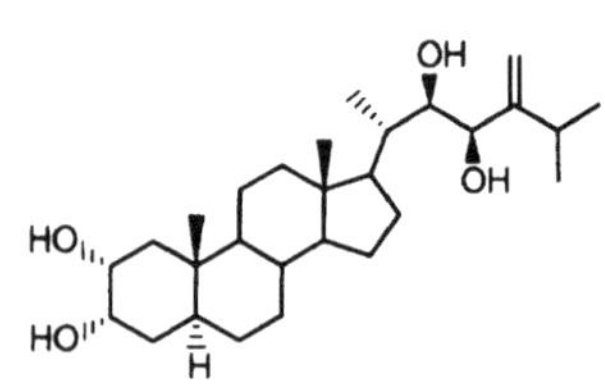

6-Deoxodolichosterone (20)

Fig. 1 *(continued)*

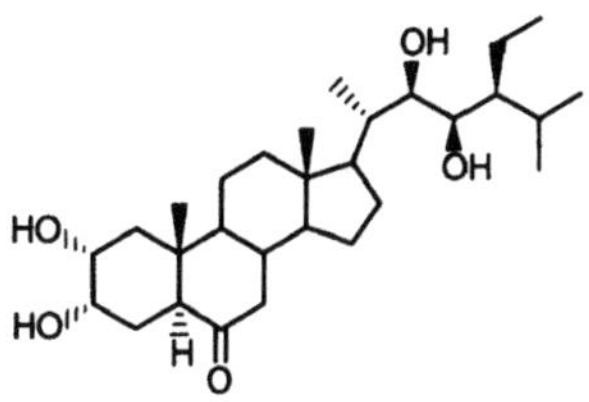

28-Homocastasterone (21)

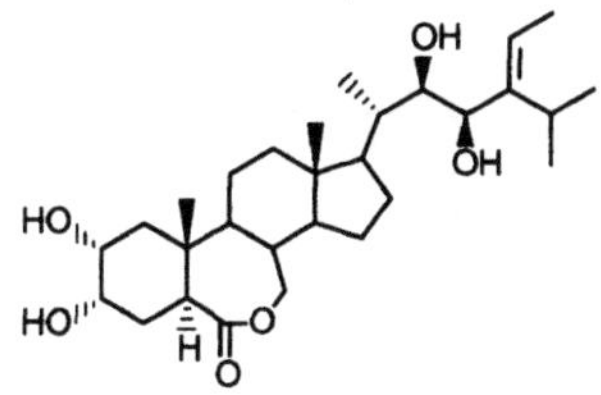

28-Homodolicholide (22)

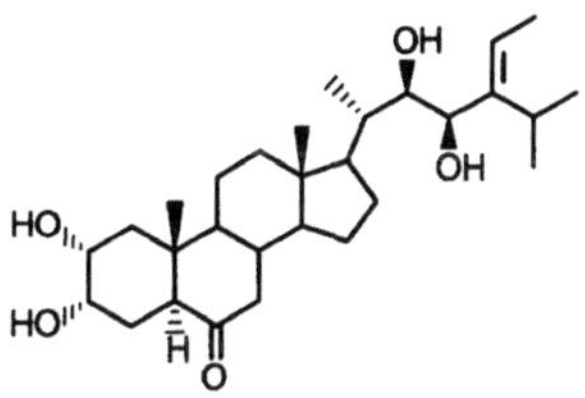

28-Homodolichosterone (23)

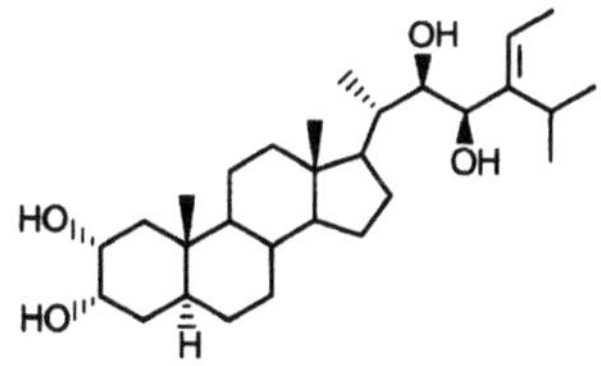

6-Deoxo-28-homodolichosterone (24)

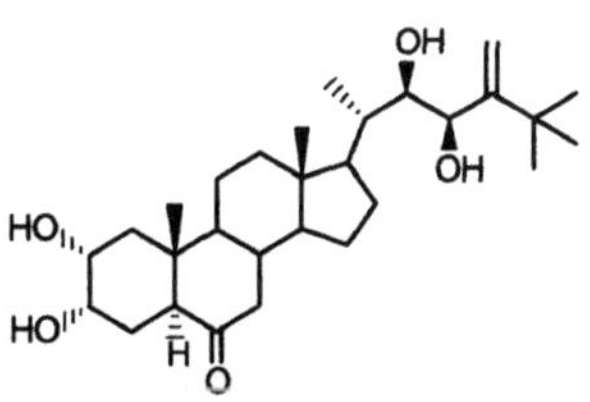

25-Methyldolichosterone (25)

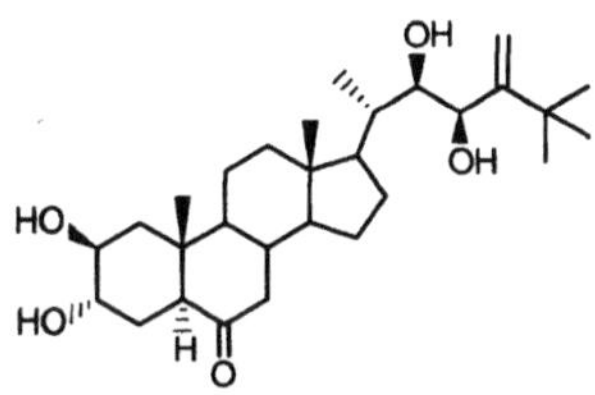

2-Epi-25-methyldolichosterone (26)

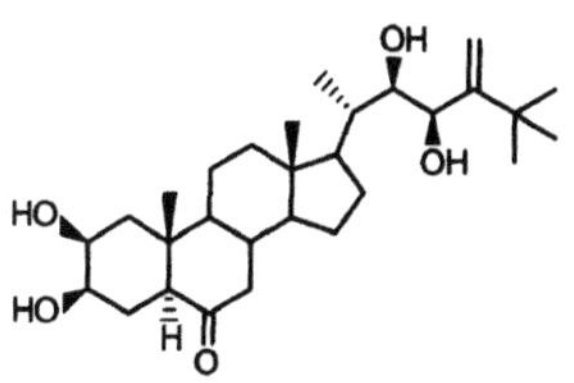

2,3-Diepi-25-methyldolichosterone (27)

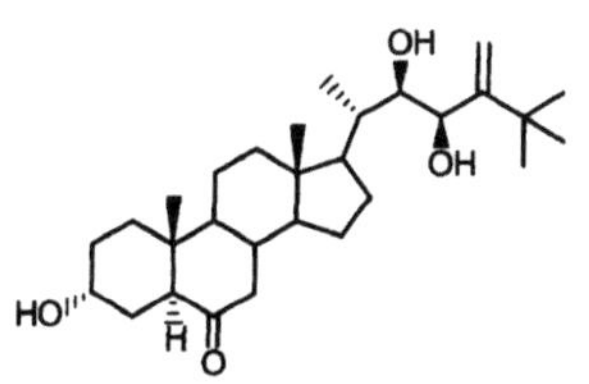

2-Deoxy-25-methyldolichosterone (28)

Fig. 1 *(continued)*

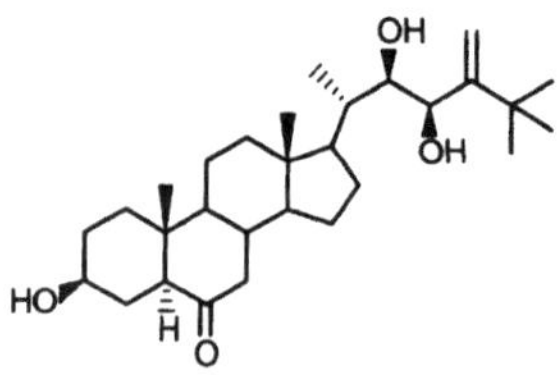

2-Epi-2-deoxy-25-methyldolichosterone (29)

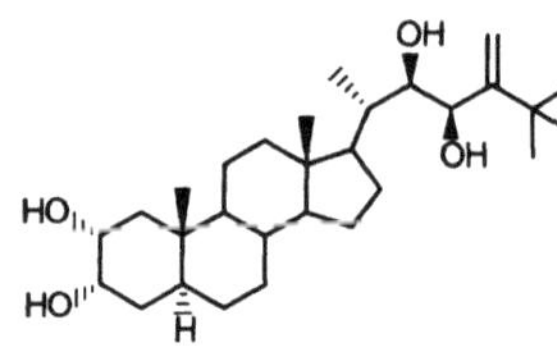

6-Deoxo-25-methyldolichosterone (30)

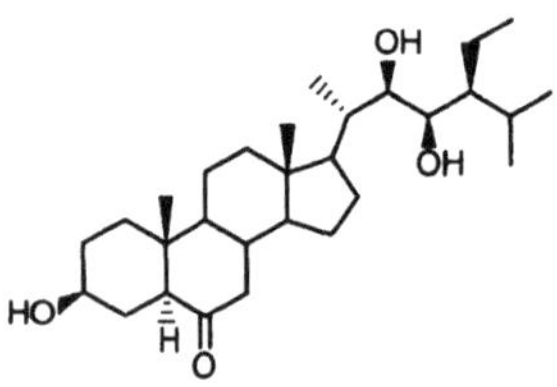

28-Homoteasterone (31)

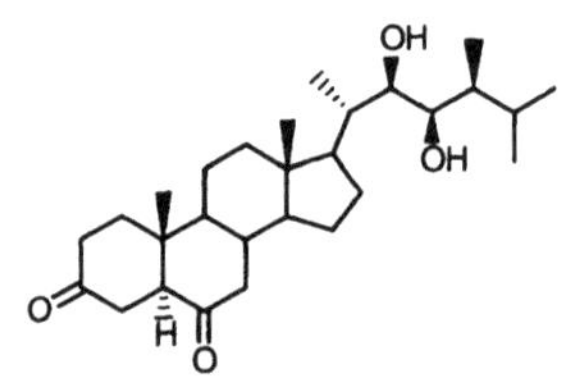

3-Dehydroteasterone (32)

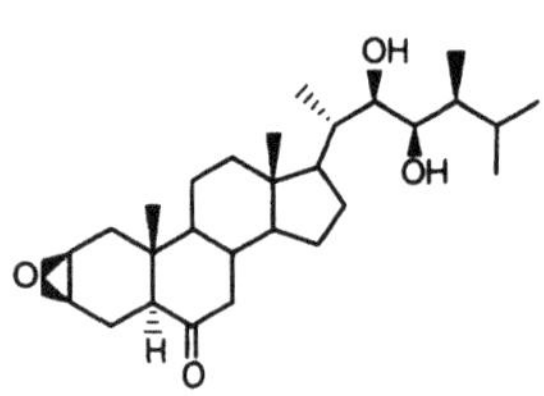

Secasterone (33)

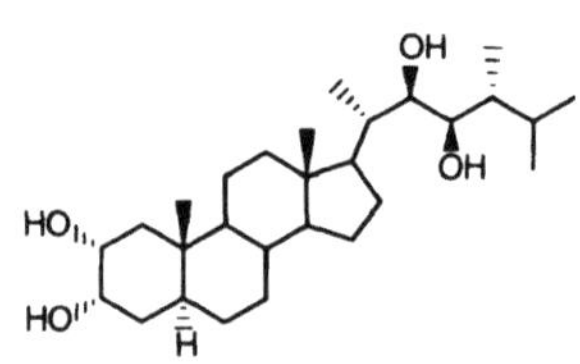

6-Deoxo-24-epicastasterone (34)

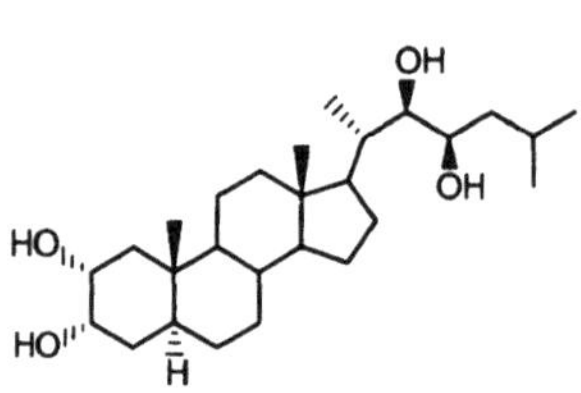

6-Deoxo-28-norcastasterone (35)

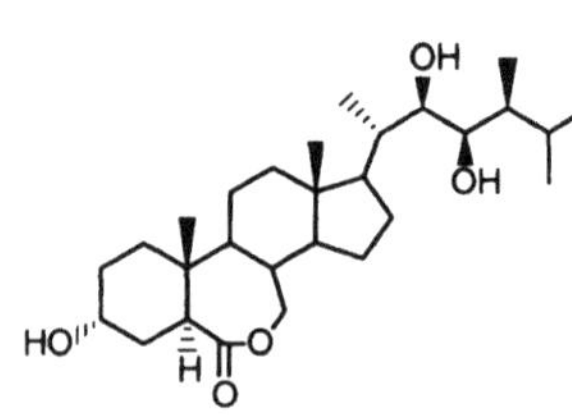

2-Deoxybrassinolide (36)

Fig. 1 *(continued)*

References, pp. 36–46

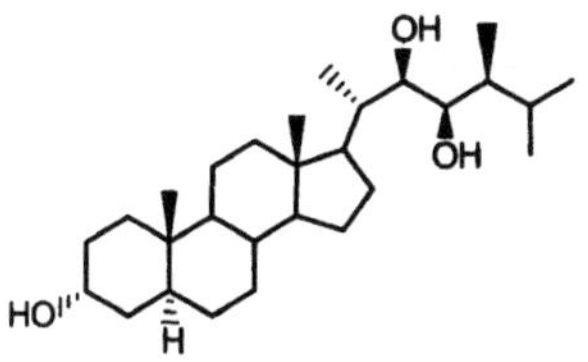

6-Deoxotyphasterol (37)

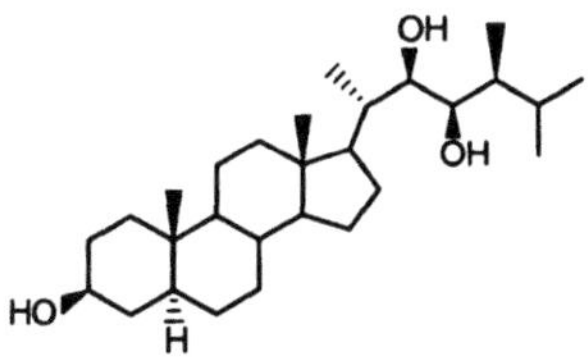

6-Deoxoteasterone (38)

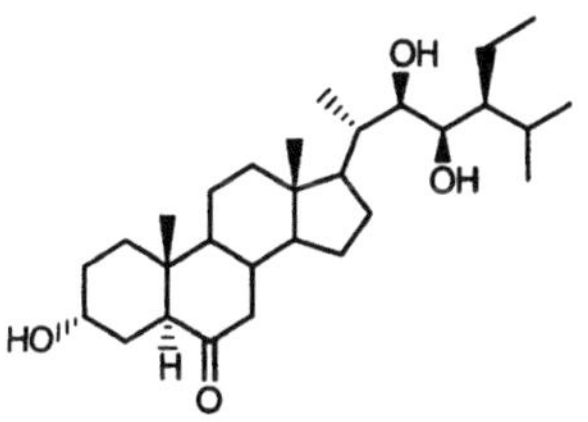

28-Homotyphasterol (39)

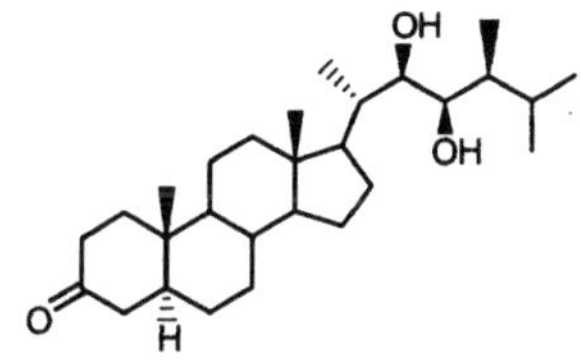

3-Dehydro-6-deoxoteasterone (40)

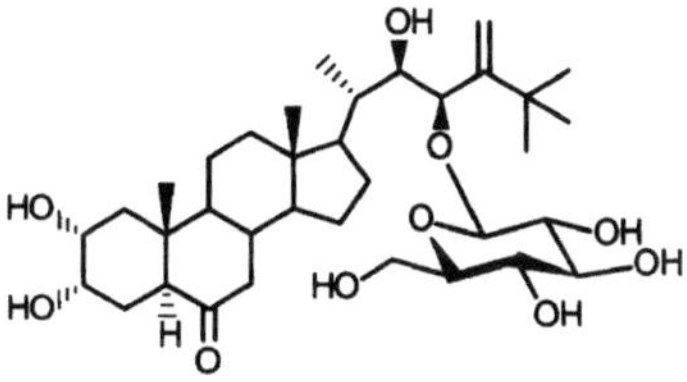

23-O-β-D-Glucopyranosyl-25-methyldolichosterone (41)

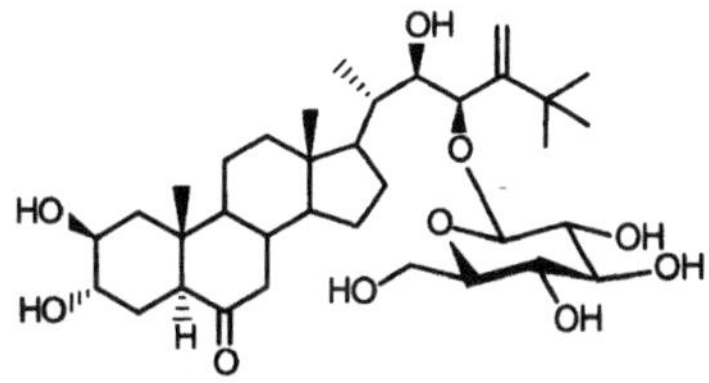

23-O-β-D-Glucopyranosyl-2-epi-25-methyldolichosterone (42)

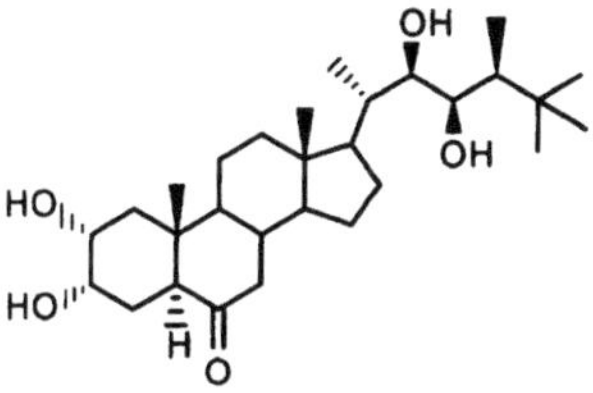

25-Methylcastasterone (43)

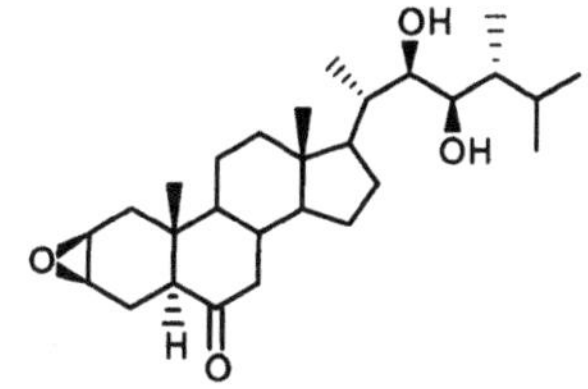

24-Episecasterone (44)

Fig. 1 *(continued)*

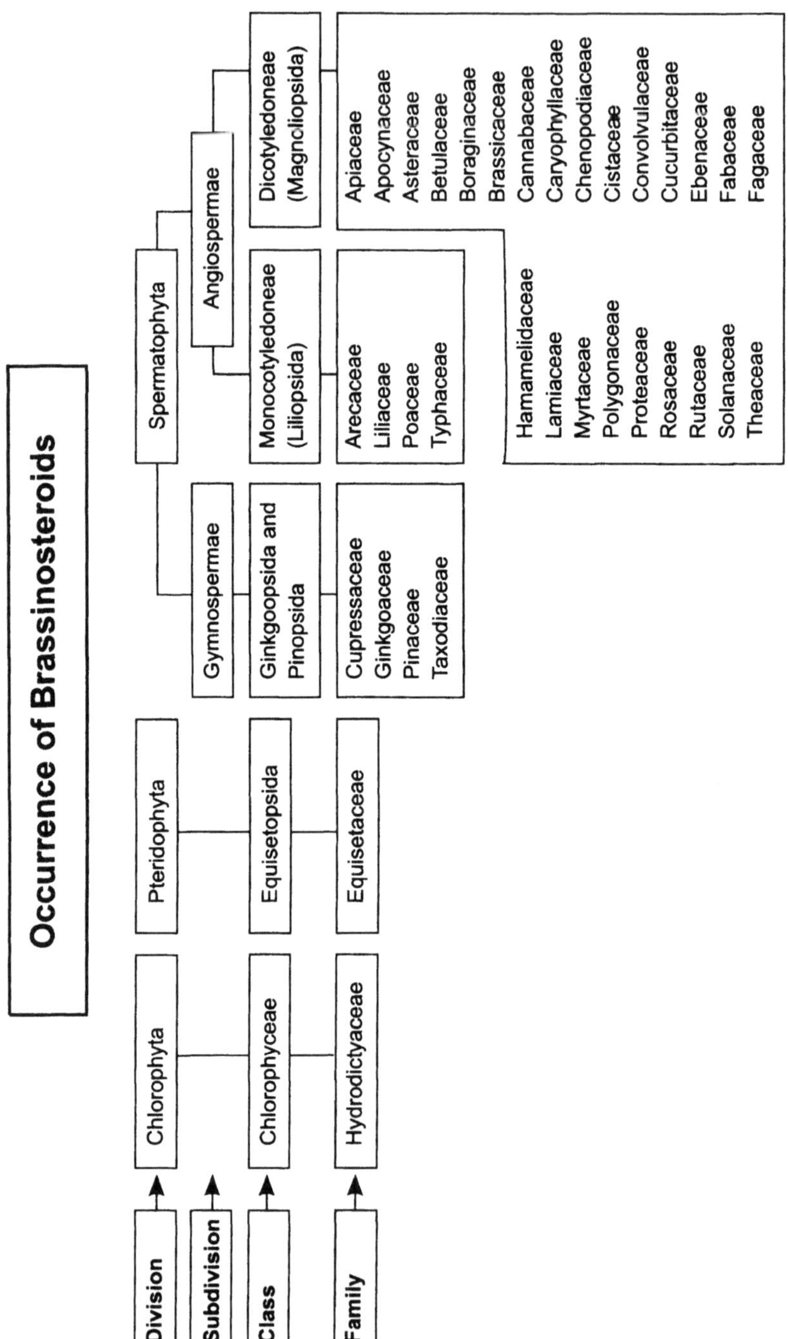

Fig. 2. Distribution of brassinosteroids in the plant kingdom

Table 1. *Natural occurrence of brassinosteroids*

Plant source	Plant family	Plant part(s)	Brassinosteroids	Ref.
Alnus glutinosa (L.) Gaertn.	Betulaceae	pollen	**1**, **3**	(*15*)
Apium graveolens L.	Apiaceae	seeds	**36**	(*16*)
Arabidopsis thaliana (L.) Heynh.	Brassicaceae	shoots	**3**, **7**, **11**, **37**	(*17*)
		seeds	**3**, **13**	(*18*)
		seeds, siliques	**1**, **3**, **7**, **11**, **37**, **38**	(*19*)
Banksia grandis Willd.	Proteaceae	pollen	**1**, **3**	(*20*)
Beta vulgaris L.	Chenopodiaceae	seeds	**3**, **14**	(*21*)
Brassica campestris var. *pekinensis* Lour.	Brassicaceae	seeds, sheath	**1**, **2**, **3**, **16**, **17**, **21**	(*22–25*)
Brassica napus L.	Brassicaceae	pollen	**1**	(*1*)
Cannabis sativa L.	Cannabaceae	seeds	**3**, **8**	(*26*, *27*)
Cassia tora L.	Fabaceae	seeds	**1**, **3**, **7**, **8**, **17**	(*28*)
Castanea crenata Sieb. et Zucc.	Fagaceae	galls	**1**, **3**, **11**, **17**	(*22*, *29–31*)
		shoots	**3**, **11**	(*30*)
Catharanthus roseus G. Don	Apocynaceae	culture cells	**1**, **3**, **7**, **8**, **11**, **37**, **38**	(*32–34*)
Cistus hirsutus Theill.	Cistaceae	pollen	**1**, **3**	(*20*)
Citrus sinensis Osbeck	Rutaceae	pollen	**1**, **3**	(*35*)
Citrus unshiu Marcov.	Rutaceae	pollen	**1**, **3**, **7**, **8**	(*20*, *36*)
Cryptomeria japonica D. Don.	Taxodiaceae	pollen, anthers	**7**, **18**, **32**	(*20*, *37*)
Cucumis sativus L.	Cucurbitaceae	seeds	**3**	(*38*)
Cupressus arizonica Greene	Cupressaceae	pollen	**1**, **3**, **7**, **8**, **11**, **19**, **21**, **32**, **37**, **40**	(*39*)
Daucus carota ssp. sativus L.	Apiaceae	seeds	**1**, **3**, **14**	(*40*)
Diospyros kaki Thunb.	Ebenaceae	seeds	**3**	(*20*, *41*)
Distylium racemosum Sieb. et. Zucc.	Hamamelidaceae	galls	**3**, **17**	(*22*)
		leaves	**1**, **3**, **16**, **17**	(*24*, *42*)
Dolichos lablab L.	Fabaceae	seeds	**1**, **3**, **11**, **18**, **19**, **20**, **22**, **23**	(*43–46*)
Echium plantagineum L.	Boraginaceae	pollen	**1**	(*20*)

Table 1 *(continued)*

Plant source	Plant family	Plant part(s)	Brassinosteroids	Ref.
Equisetum arvense L.	Equisetaceae	strobilus	**3**, **16**, **17**, **19**	*(47)*
Eriobotrya japonica Lindl.	Rosaceae	flower buds	**3**	*(20)*
Erythronium japonicum Decne	Liliaceae	pollen, anthers	**7**	*(48)*
Eucalyptus calophylla R. Br.	Myrtaceae	pollen	**1**	*(20)*
Eucalyptus marginata Sn.	Myrtaceae	pollen	**19**	*(20)*
Fagopyrum esculentum Moench	Polygonaceae	pollen	**1**, **3**	*(49)*
Ginkgo biloba L.	Ginkgoaceae	seeds	**8**	*(50)*
Gypsophila perfoliata L.	Caryophyllaceae	seeds	**13**	*(51)*
Helianthus annuus L.	Asteraceae	pollen	**1**, **3**, **17**	*(52)*
Hydrodictyon reticulatum (L.) Lagerheim	Hydrodictyaceae	green alga	**14**, **21**	*(53)*
Lilium elegans Thunb.	Liliaceae	pollen	**1**, **3**, **7**, **8**	*(54)*
Lilium longiflorum Thunb.	Liliaceae	anthers	**1**, **3**, **7**, **9**, **10**, **32**	*(36, 42, 55)*
		pollen	**9**, **10**	*(56)*
Lolium perenne L.	Poaceae	pollen	**43**	*(57)*
Lycopersicon esculentum Mill.	Solanaceae	shoots	**3**, **11**, **17**	*(58)*
Lychnis viscaria L.	Caryophyllaceae	seeds	**14**, **44**	*(59)*
Ornithopus sativus Brot.	Fabaceae	seeds	**3**, **14**	*(60)*
		shoots	**3**, **11**, **14**, **34**, **35**	*(61)*
Oryza sativa L.	Poaceae	shoots	**1**, **3**, **19**	*(62)*
		bran	**11**, **31**, **39**	*(63)*
		seeds	**3**, **8**, **11**	*(64)*
Perilla frutescens Britton.	Lamiaceae	seeds	**3**	*(64)*
Phalaris canariensis L.	Poaceae	seeds	**3**, **8**	*(65)*
Pharbitis purpurea Voigt	Convolvulaceae	seeds	**3**, **17**	*(66)*
Phaseolus vulgaris L.	Fabaceae	seeds	**1**, **3–8**, **11**, **12**, **15**, **18–20**, **24–30**, **41**, **42**	*(67–72)*
Phoenix dactylifera L.	Arecaceae	pollen grains	**14**	*(73)*

References, pp. 36–46

Picea sitchensis Trautv. ex Mey	Pinaceae	shoots	**3**, **7**	(*74*)
Pinus sylvestris Lour.	Pinaceae	cambial region	**1**, **3**	(*75*)
Pinus thunbergii Parl.	Pinaceae	pollen	**3**, **7**	(*76*)
Pisum sativum L.	Fabaceae	seeds	**1**, **3**, **7**, **11**, **36**	(*77*)
		shoots	**1**, **3**, **11**	(*78*)
Psophocarpus tetragonolobus DC.	Fabaceae	seeds	**1**, **3**, **11**, **21**	(*20*)
Raphanus sativus L.	Brassicaceae	seeds	**1**, **3**, **8**, **31**	(*79*, *80*)
Rheum rhabarbarum L.	Polygonaceae	panicles	**1**, **3**, **14**	(*81*)
Robinia pseudo-acacia Oswald	Fabaceae	pollen	**3**, **7**, **11**	(*82*)
Secale cereale L.	Poaceae	seeds	**3**, **7**, **8**, **11**, **17**, **21**, **33**	(*83*)
Solidago altissima L.	Asteraceae	stem	**1**	(*20*)
Thea sinensis L.	Theaceae	leaves	**1**, **3**, **7**, **8**, **17**, **21**	(*22*, *25*, *84*, *85*)
Triticum aestivum L.	Poaceae	grain	**3**, **7**, **8**, **11**, **32**	(*86*)
Tulipa gesneriana L.	Liliaceae	pollen	**7**	(*20*, *36*)
Typha latifolia G. F. W. Mey	Typhaceae	pollen	**7**	(*87*)
Vicia faba L.	Fabaceae	seeds	**1**, **3**	(*88*)
		pollen	**1**, **3**, **13**, **17**, **19**	(*88*, *89*)
Zea mays L.	Poaceae	pollen	**3**, **7**, **8**, **17**, **21**	(*91*, *92*)
		seeds	**3**, **8**, **11**, **17**	(*93*)

Furthermore, natural occurrences of brassinosteroids are also known from a pteridophyte (*Equisetum arvense*) and a chlorophyte, the alga *Hydrodictyon reticulatum* (Table 1). Brassinosteroids were detected in several organs such as stems, leaves, pollen and seeds. Pollen and immature seeds are especially rich sources of brassinosteroids. Till now they are not known to occur in roots. Compared with other plant hormones brassinosteroid concentrations in vegetative tissues are normally very low, ranging in the ng to μg level per kg plant material.

Castasterone (**3**) is the most widely distributed brassinosteroid followed by brassinolide (**1**). While the trihydroxybrassinosteroids typhasterol (**7**) and teasterone (**8**) as well as 6-deoxocastasterone (**11**) and 28-norcastasterone (**17**) are also widely distributed, compounds of the 24-*epi* series such as 24-epibrassinolide (**13**) and 24-epicastasterone (**14**) represent relatively scarce brassinosteroids (Table 1).

3. Structures

Brassinosteroids can be derived from the 5α-cholestane skeleton and their structural variations comprise the substitution pattern at rings A and B as well as the C-17 side chain (*3, 8–13*).

A general structural feature of the most brassinosteroids are the two vicinal hydroxy groups at ring A (C-2α and C-3α), *e.g.* in brassinolide (**1**) and castasterone (**3**). Furthermore, also the other ring A epimers (2α,3β; 2β,3α; 2β,3β) were found (**4, 5, 6, 12, 15, 26, 27**) (*67*). Some biogenetic precursors of brassinolide like typhasterol (**7**) and teasterone (**8**) have only one hydroxy function in ring A (C-3α or β). Brassinosteroids with an epoxide in ring A, secasterone (**33**) (*83*) and 24-episecasterone (**44**) (*59*), were also found.

Ring B is characterized either by a seven-membered 6a-oxalactone (*e.g.* in brassinolide, (**1**)) or is a normal 6-membered ring as a 6-oxo derivative (*e.g.* castasterone, **3**) or a 6-deoxo type (*e.g.* 6-deoxocastasterone, **11**).

The observed variations in the side chain moiety reflect the corresponding biogenetic precursor phytosterols. Based on the cholestane side chain, the various skeletons can thus be grouped into seven categories with different substituent at C-24 and C-25, respectively:

1) 24*S*-methyl (**1, 3, 4, 5, 6, 7, 8, 9, 10, 11, 12, 32, 33, 36, 37, 38, 40**)
2) 24*R*-methyl (**13, 14, 15, 34, 44**)
3) 24-methylene (**18, 19, 20**)
4) 24*S*-ethyl (**2, 21, 31, 39**)
5) 24-ethylidene (**22, 23, 24**)

6) 24-methylene-25-methyl (**25**, **26**, **27**, **28**, **29**, **30**, **41**, **42**) and 24*S*-methyl-25-methyl (**43**)
7) without substituent at C-24 (**16**, **17**, **35**)

Except for some not yet fully characterized brassinosteroids with an oxo function at C-23 (*37*) all brassinosteroids described so far have a vicinal *R,R*-configurated diol functionality at C-22 and C-23.

In addition to the free brassinosteroids also some sugar and fatty acid conjugates have been detected. Thus, 23-O-β-D-glucopyranosyl-25-methyldolichosterone (**41**) and 23-O-β-D-glucopyranosyl-2-epi-25-methyldolichosterone (**42**) were found in immature seeds of *Phaseolus vulgaris* (*68*) while teasterone-3-laurate (**9**) and teasterone-3-myristate (**10**) occur in the anthers of *Lilium longiflorum* (*55*, *56*).

4. Isolation and Purification

The extraction and purification of brassinosteroids involves a sequence of partition and chromatographic processes. Brassinosteroids can be extracted from the plant material with methanol or methanol/ethylacetate followed by partition between water/chloroform and 80% methanol/*n*-hexane. The brassinosteroid containing 80% methanol fraction is further purified by successive chromatography on silica gel, Sephadex LH-20 and ion exchange on DEA followed by preparative HPLC. There are some variations with respect to extraction and chromatography. In all cases it is necessary to use a sensitive bioassay for monitoring the brassinosteroid containing fractions during the chromatographic steps. The most useful bioassay for brassinosteroids is the rice-lamina inclination test (*4*, *94*, *95*) which was originally developed for auxins. Using this assay one can detect 0.1 nM brassinolide (**1**) and below. Less sensitive tests are the bean second internode assay (*1*) and the wheat leaf-unrolling test (*96*). The application of a radioimmunoassay has been also described (*97*). A common isolation and purification procedure of brassinosteroids is given in Scheme 1 (*7*).

5. Analysis of Brassinosteroids

5.1. GC-MS Analysis

The low amounts of brassinosteroids in plant material demand very sensitive analytical methods for the structural identification of

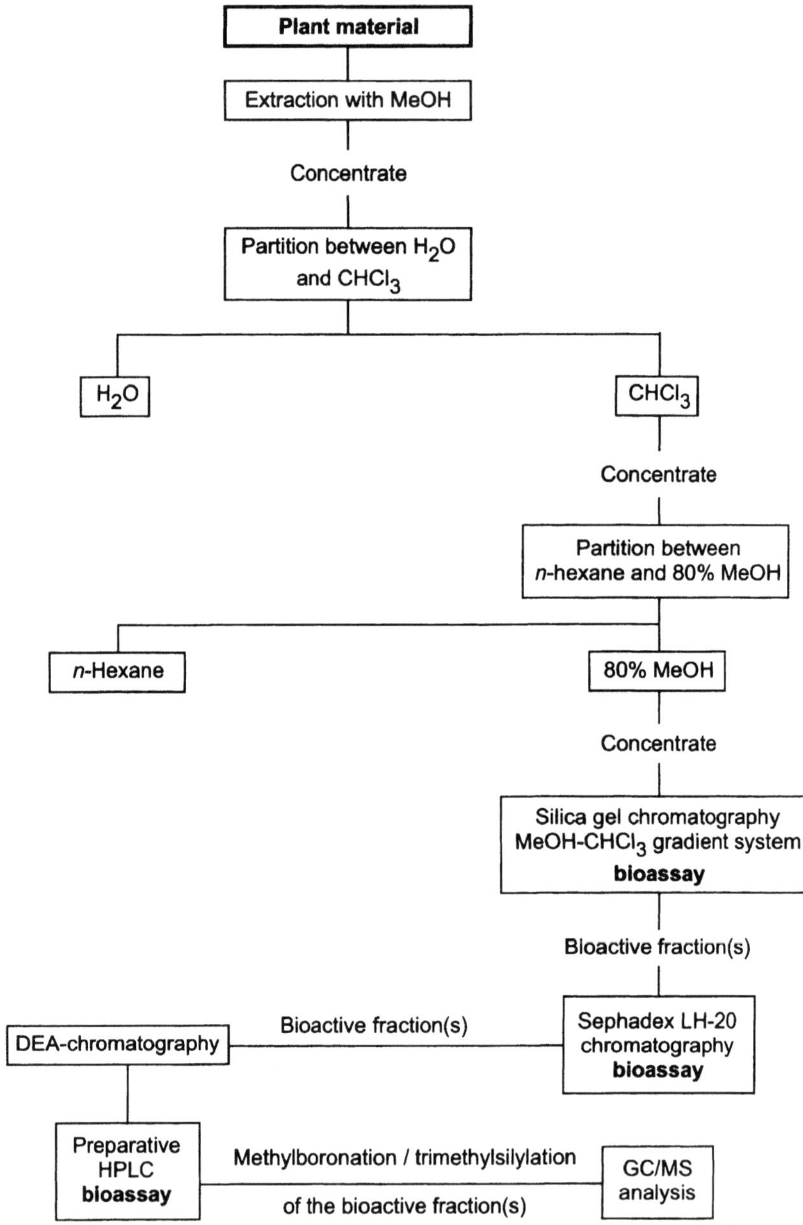

Scheme 1. Isolation and purification of brassinosteroids

brassinosteroids. The most used microanalytical method for analyzing brassinosteroids is combined gas chromatography-mass spectrometry (GC-MS). For this analysis the brassinosteroids have to be converted into volatile derivatives. In case of tetrahydroxylated brassinosteroids such as brassinolide (**1**) or castasterone (**3**) the bismethylboronation (BMB) of the vicinal hydroxy functions using methylboronic acid and pyridine is the method of choice (*22*, *98*, *99*). The trihydroxylated brassinosteroids such as typhasterol (**7**) and teasterone (**8**) can be derivatized by methylboronation of the side chain followed by trimethylsilylation of the C-3-hydroxy group (*98*). The electron impact (EI) mass spectra of the bismethylboronates (BMB) and brassinosteroid methylboronate-trimethylsilyl (MB-TMS) derivatives yield a much information about the brassinosteroid structure. They show a molecular ion and very characteristic key ions, thus allowing assignment both of the side chain structure and the basic steroidal skeleton (*20*, *98*, *99*).

The main fragmentation of brassinosteroids is characterized by side chain cleavages. The key fragments can be classified into four different types (Schemes 2–5):

1) fragments characterizing the side chain,
2) fragments comprising the steroid skeleton with rings ABCD,
3) fragments obtained by cleavages in rings B, C and D,
4) fragments obtained by fragmentation of rings AB.

The key ion of type **a** formed by an α-cleavage between C-20/C-22 is typical for brassinosteroids with a saturated side chain (Scheme 2). Methylation at C-24 is indicated by an ion at *m/z* 155 (e. g. **1**, **3**, **7**, **33**, **40**). A mass shift to *m/z* 169 indicates an additional CH_2-unit in the side chain and gives a hint to 28-homobrassinosteroids such as **2**, **21**, **31** and **39**. Consequently, the ion at *m/z* 141 represents a key ion of the 28-norbrassinosteroids **16**, **17** and **35**. A side chain double bond between C-24/C-28, as in case of dolichosterone (**19**) and 28-homodolichosterone (**23**), leads to a shift of 2 mass units towards lower masses (ion **b**, *m/z* 153 and 167, respectively) compared with the corresponding saturated ones. This double bond triggers a further fragmentation leading to an ion of type **c** (*m/z* 124 and 138, respectively), which can suffer further fragmentation to give the ion (**c**-$CH_3B{=}O$) at *m/z* 82 (C_{28}-brassinosteroids) and *m/z* 96 (C_{29}-brassinosteroids). Differentiation between 28-homodolichosterone (**23**) and 25-methyldolichosterone (**25**) is possible by the appearance of a significant ion at *m/z* 123 originating by loss of a methyl from **c** via allylic cleavage (*72*). Recently, four isomers of ξ-epi-23-dehydrobrassinolide with undefined stereochemistry were described

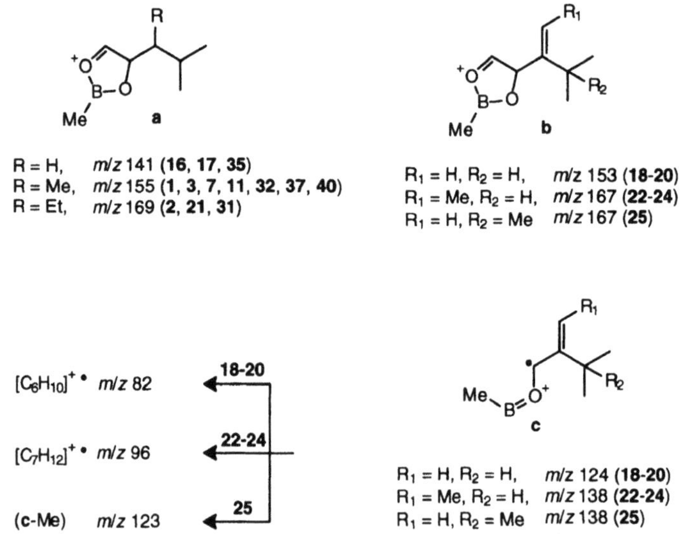

Scheme 2. Key ions resulting from the side chain

(*37*). In that cases the methylboronation of the side chain leads to two unusual bismethylboronate derivatives involving two different enol forms. While one product shows a side chain fragment at *m/z* 181 (cleavage C-17/C-20), the second one display formally a **b**-type ion (*m/z* 153) (*37*).

Cleavages in the side chain also lead to ions comprising the rings ABCD and the C-atoms 20 and 21 (Scheme 3). The formation of the **d**-type ion, being complementary to the ions of type **a** and **b**, respectively, can be accompanied by a hydrogen transfer in each direction. While in the case of the saturated derivatives (*e.g.* of **1** and **3**) the ions thus formed appear at (**d**+H), the unsaturated ones (*e.g.* of **18** and **19**) do not show a hydrogen transfer. In 25-methyldolichosterone (**25**) an ion at *m/z* 356 (**d**-H) is first formed. The fission C-17/C-20 (loss of the side chain) leads to ion **e** representing the ABCD rings. In case of brassinosteroids with an unsaturated side chain this cleavage is accompanied by a double hydrogen transfer leading to an ion of type (**d**-2H) (*99*). The saturated brassinosteroids show another key ion, comprising the rings ABCD and a part of the side chain (C-20 to C-23), which results from an α-cleavage of the bond C-23/C-24. The thus formed fragments at $[M-57]^+$(*e.g.* of **17**), $[M-71]^+$ (*e.g.* of **3**) and $[M-85]^+$-ions (*e.g.* of **21**),

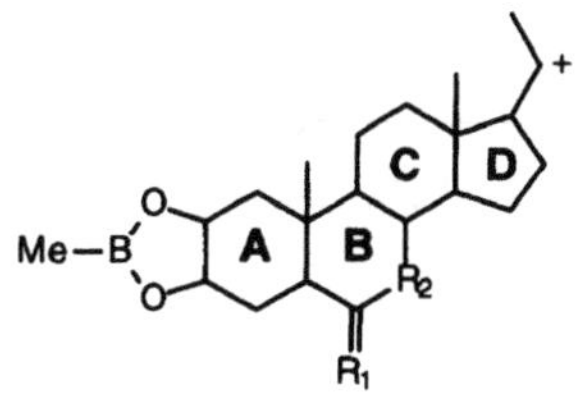

d

$R_1 = H_2$, $R_2 = CH_2$, *m/z* 343 (**11**, **12**, **17**)
$R_1 = O$, $R_2 = CH_2$, +H *m/z* 358 (**3**, **17**, **19**)
-H *m/z* 356 (**25**)
$R_1 = O$, $R_2 = OCH_2$ +H *m/z* 374 (**1**, **2**, **18**)

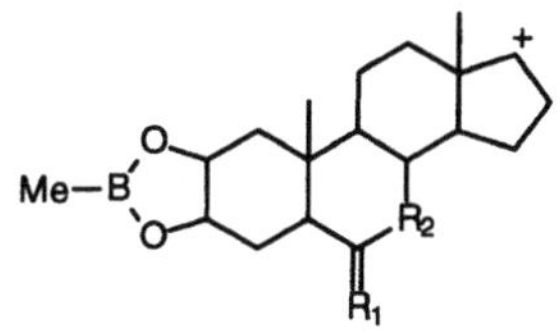

e

$R_1 = H_2$, $R_2 = CH_2$, -2H *m/z* 313 (**11**, **17**)
$R_1 = O$, $R_2 = CH_2$, *m/z* 329 (**3**, **17**, **19**)
-2H *m/z* 327 (**25**)
$R_1 = O$, $R_2 = OCH_2$ *m/z* 345 (**1**, **2**, **16**, **18**)

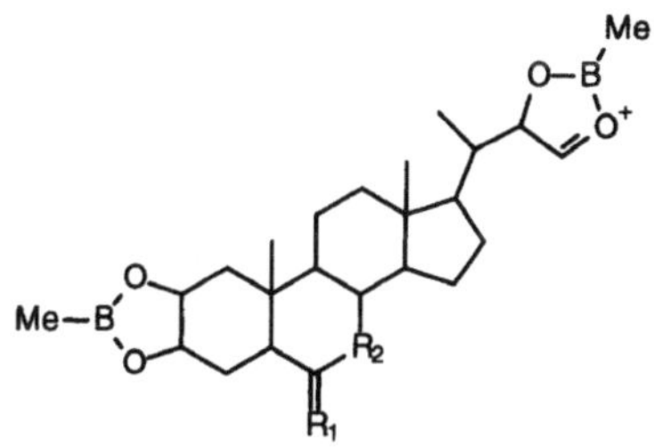

$[M-57]^+$ (**16**, **17**, **35**)
$[M-71]^+$ (**1**, **3**, **11**)
$[M-85]^+$ (**2**, **21**)

$R_1 = H_2$, $R_2 = CH_2$
$R_1 = O$, $R_2 = CH_2$
$R_1 = O$, $R_2 = OCH_2$

Scheme 3. Key ions comprising the steroid skeleton

respectively, give a hint concerning the substitution at C-24 (Scheme 3). In addition brassinosteroids with a double bond C-24/C-28 (*e.g.* **18** and **19**) display a prominent $[M-43]^+$-ion (loss of C_3H_7) as a result of cleavage between C-24 and C-25 (*99*).

Another type of fragment, originating in cleavages in the rings B, C and D, yields further possibilities for differentiating between the several types of brassinosteroids. Thus, a significant ion comprising the rings ABC is ion **f** which results from a hydrogen transfer from the charged species (Scheme 4). The **f**-type ion is typical of both 6-deoxo and 6-oxobrassinosteroids with two hydroxy groups at ring A (*m/z* 273 and

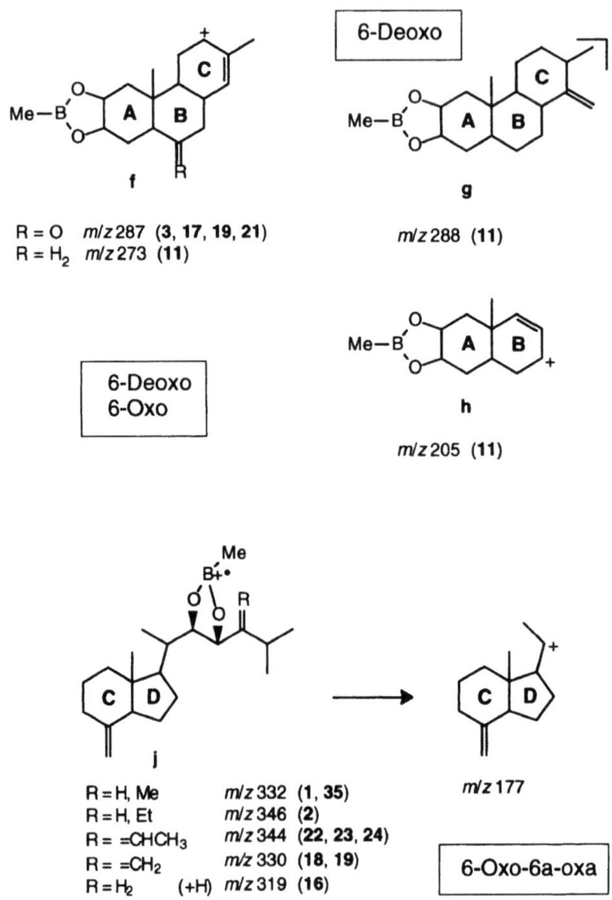

Scheme 4. Key ions comprising rings ABC and CD

m/z 287, respectively). In case of the 6-deoxo compounds ion **g** (*m/z* 288) is also of importance. Ion **h** (*m/z* 205) obtained by cleavage of ring C is a significant peak in the EI mass spectra of 6-deoxobrassinosteroid-BMB derivatives (*25*). An important key ion, characteristic of lactone type brassinosteroids such as brassinolide (**1**), is an ion **j** formed by cleavage of ring B between C-9/C-10 and O/C-7. It shows a mass shift corresponding to the side chain structure (*99*). Therefore, the ion at *m/z* 177, containing rings CD and C-20, C-21, also indicates the lactone type (Scheme 4).

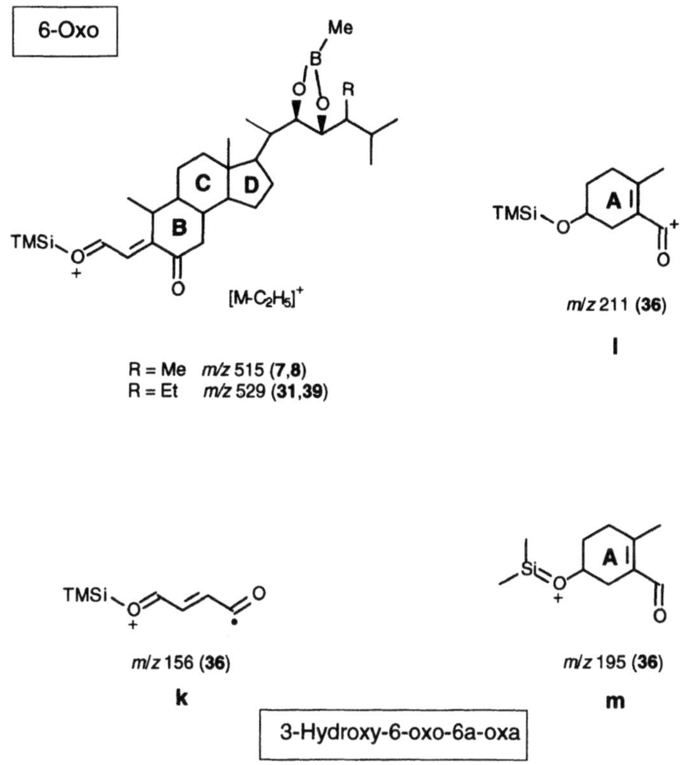

Scheme 5. Key ions resulting from ring AB fragmentation of 2-deoxybrassinosteroids

A typical fragmentation involving rings AB is only observed in the EI mass spectra of the methylboronate-trimethylsilylether (MB-TMS) of the trihydroxybrassinosteroids **7**, **8**, **31**, **36**, **39**. The MB-TMS derivatives of typhasterol (**7**), teasterone (**8**), 28-homotyphasterol (**39**) and 28-homoteasterone (**31**) display, besides significant $[M-Me]^+$- and $[M-TMSiOH]^+$ ions, a prominent $[M-C_2H_5]^+$-ion. Its formation is initiated by an α-cleavage between C-2 and C-3 and includes the loss of the carbons C-1 and C-2, respectively, connected with a hydrogen transfer (Scheme 5). A specific key ion, also yielding evidence for the presence of a 3-hydroxy-6a-oxalactone structure of 2-deoxybrassinolide (**36**), is ion **k** (*m/z* 156). This conjugated oxonium ion is formed by cleavages through rings A and B and was established by high-resolution mass measurements (*100*). The same is true for the ions at *m/z* 211 (**l**) and 195 (**m**). Ion **m** probably originates from the [M-Me]$^+$-ion. In case of the MB-TMS derivatives of the 3-hydroxy-6-deoxobrassinosteroids **37** and **38** ions of type (**f**-TMSiOH) at *m/z* 215 (base peak) and (**g**-TMSiOH) at *m/z* 230 (*38*) are formed. These ions are also typical for the EI mass spectra of ring ABCD saturated phytosterol acetates and their TMS derivatives, respectively (*101*).

Compounds with an oxo- or epoxy-function in ring A such as 3-dehydroteasterone (**32**), secasterone (**33**) and 3-dehydro-6-deoxoteasterone (**40**) form methylboronates (MB). Generally, they show the same side chain fragmentation (ion **a**) as other 6-oxo and deoxobrassinosteroids, respectively (Scheme 2). The ions **d**, **e**, **f** and **g**, comprising the steroidal ring system as [M-71]$^+$, appear with the corresponding mass shift (Schemes 3 and 4) (*7*, *38*, *41*, *83*, *86*).

In summary one can conclude that the EI mass spectra of derivatized brassinosteroids (BMB, MB and MB-TMS) yield very valuable structural information. There is a set of ions characteristic for the substructures of the single brassinosteroid (Schemes 2–5). Combining the several types of fragments it is possible to deduce the general basic brassinosteroid structure. Furthermore, the derivatized brassinosteroids exhibit sharp peaks in capillary GC and most of them can be separated from each other, although some isomers at C-22 and C-23 can have very close retention times (*99*). For reproducible retention data of the brassinosteroids both the Kovats retention indices using a hydrocarbon mixture (*77*) as well as the relative retention time with respect to an internal standard, *e.g.* 5α-cholestane or castasterone (*61*, *83*), can be used. Additionally, considering all results from the gas chromatographic behaviour one can deduce the rule that as regards the 24-epimers generally the members of the 24*S* series elute earlier than the 24*R* ones (*7*, *16*, *60*, *61*, *83*, *99*).

Using only the molecular ion and the key ions discussed above the method of GC-selected ion monitoring (SIM) can successfully be applied for identification of brassinosteroids (*20*, *22–24*, *98*, *99*). Impressive examples which illustrate the application of GC-high-resolution (HR)-MS-SIM are the identification of brassinosteroids from pollen of *Fagopyrum esculentum* (*49*) and *Vicia faba* (*89*). This technique is capable for detecting brassinosteroids at the picogram level.

Combined gas chromatography-chemical ionization mass spectrometry (GC-CIMS) yields not only prominent [M+H]$^+$ ions, but also provides improved sensitivity (*98*, *99*). The selectivity of MS/MS techniques (daughter ion scan) was demonstrated in the identification of brassinolide (**1**) and castasterone (**3**) from *Alnus glutinosa* by GC-CIMS/MS (*15*).

5.2. HPLC Analysis

GAMOH *et al.* have published an application of HPLC combined with UV (*102*) and fluorimetric detection (*90*, *92*, *103*, *104*) for determining brassinosteroids by using different boronic acid derivatization reagents. In these cases the brassinosteroids were boronated with 9-phenanthreneboronic acid (*90*), dansylaminophenylboronic acid (*92*), 1-naphthaleneboronic acid (*102*), 1-cyanoisoindole-2-*m*-phenylboronic acid (*103*) and *m*-aminophenylboronic acid (*104*). The detection is possible at levels of 20–50 pg. Dansylaminophenylboronic acid seems to be the most promising reagent because the brassinosteroids can be detected at longer wavelengths (Ex 345 nm/Em 515 nm) compared to the other derivatives. Another micro-scale method for the determination of brassinosteroids using ferroceneboronates represents HPLC combined with electrochemical detectors (*105*).

Recently, HPLC/MS techniques were used for structural investigations on brassinosteroids. While combined HPLC-*frit* FABMS (fast atom bombardment mass spectrometry) using negative ionization was firstly applied in the structure elucidation of teasterone 3-O-myristate (**10**) from *Lilium longiflorum* anthers (*55*), the HPLC-APCIMS (atmospheric pressure chemical ionization mass spectrometry) was described as an useful tool for identification of brassinosteroids (*27*, *106*).

5.3. NMR Spectroscopy

Detailed NMR studies of native and synthetic brassinosteroids including 2D-experiments (H,H-COSY, C,H-COSY, HMBC, HMQC) were carried out to provide assignments for biologically important brassinosteroids (*107*, *108*). NMR-spectroscopy was shown also to be a powerful method in the structure elucidation of brassinosteroid metabolites (*7*).

6. Synthesis

The unique structure of brassinosteroids and the requirement for analytical standards as well as sufficient amounts for biological studies have greatly stimulated synthetic efforts. Already a short time after the discovery of brassinolide (**1**) three different synthetic pathways leading to **1** were published by American and Japanese chemists (*109–111*).

Nearly all syntheses of brassinosteroids are partial syntheses, starting mostly from readily available Δ^{22}-unsaturated sterols like stigmasterol (**45**), ergosterol (**51**) and campesterol (**54**) or even the scarce 22-dehydrocampesterol (crinosterol). Also suitable phytosterol mixtures, e.g. from oil of soy bean or rape, have been used. Whereas the typical structural features of the A/B rings can be introduced by classical reactions of steroid chemistry, the construction of the more original side chain moiety involves either a C_{22}-aldehyde as key intermediate or direct use of the Δ^{22} double bond of a suitable sterol precursor for introduction of the desired diol functions is adopted. Also other steroids, e.g. cholic acids or 20-keto pregnanes, have been used as starting compounds in some cases. For structure-activity relationships a large number of structural analogs of the native hormones have been prepared. The intensive development in brassinosteroid synthesis, which is still rapidly going on, has been subject of a series of reviews (*112–117*). A comprehensive treatment of this topic including synthesis of numerous structural analogs is given in ref. (*13*). In the following two typical examples of practically useful synthesis, leading to the most important native members brassinolide (**1**) and 24-epibrassinolide (**13**), are described briefly.

Because of its high biological activity brassinolide (**1**) represents a preferred target for synthetic efforts, and about twenty different routes to it have been published up to now. From the numerous published methods in the following an improved synthesis by WATANABE *et al.* (*118*) as one of the most practicable and effective procedures is demonstrated (Scheme 6). In this sequence the known 22-aldehyde (**46**), readily obtainable in 5 steps and 70% yield from commercial stigmasterol (**45**) (*119*), was reacted with (*Z*)-prop-1-enylmagnesium bromide to give preferentially the (22*S*,23*Z*)-23-en-22-ol **47**. Orthoester Claisen rearrangement of **47** afforded smoothly ester **48**. Its ethoxycarbonyl function was transformed via $LiAlH_4$-reduction to the corresponding alcohol and subsequent reductive elimination of its tosylate to give compound **49** with the desired crinosterol side chain. Acid catalyzed deprotection of **49** followed by isomerization of the resulting 3,5-cycloketone with pyridinium toluene-p-sulfonate and LiBr in N,N-dimethylacetamide yielded 2,22-dien-6-one **50**. For the subsequent introduction of both vicinal functions in **50** asymmetric dihydroxylation with OsO_4 is the common procedure, leading stereoselectively not only to the 2α,3α-diol, but in the presence of the Sharpless catalyst 1,4-bis(9-O-dihydroquinidinyl)phthalazine (*120–121*) as chiral ligand preferentially to castasterone (**3**) with the desired (22*R*,23*R*) configuration (22*R*,23*R*:22*S*,23*S* ratio = 9:1). In such a manner compound **3** was obtained from

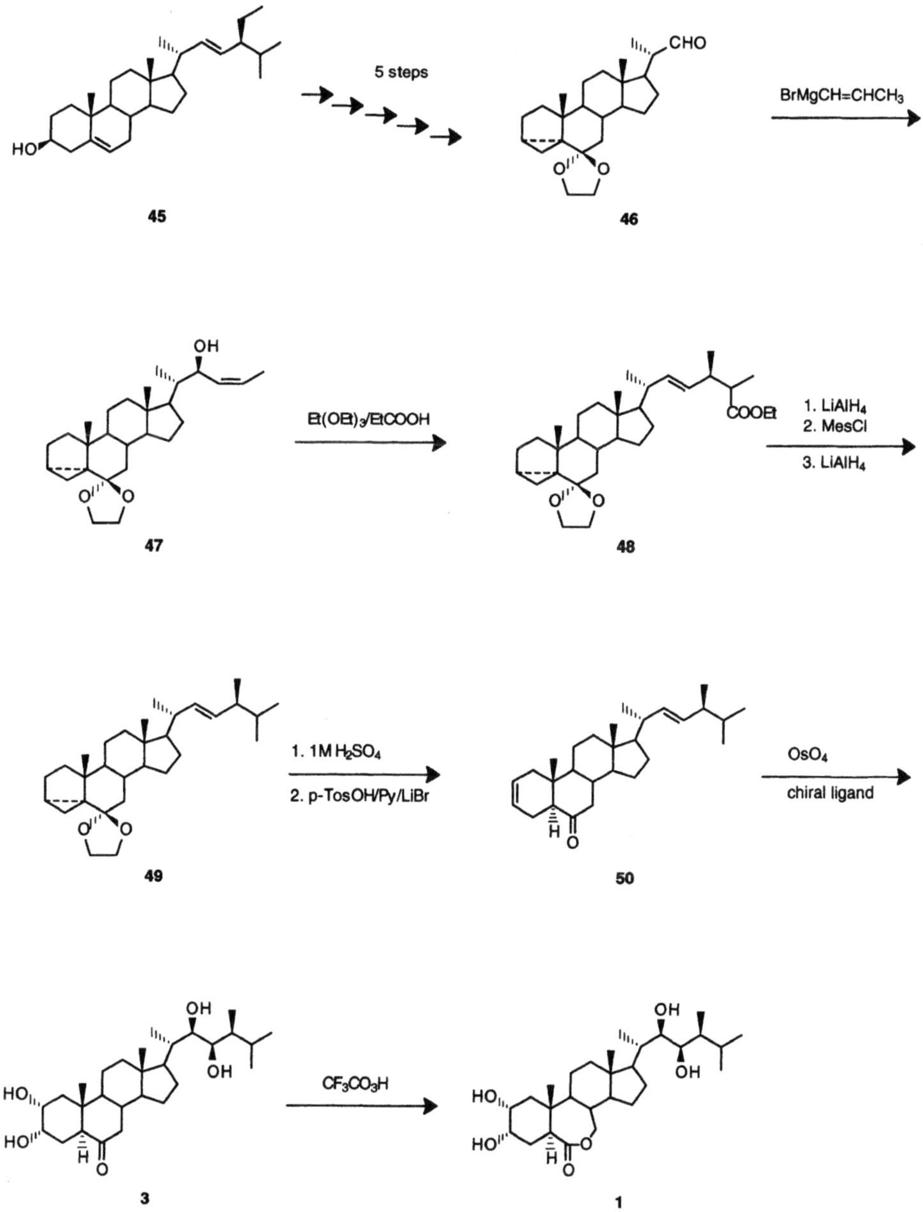

Scheme 6. Synthesis of brassinolide (**1**) by WATANABE *et al.* (1997)

stigmasterol (**45**) in 22% overall yield by 13 steps. Direct Baeyer-Villiger oxidation of **3** with trifluoroperoxy acetic acid as useful standard reagent affords smoothly brassinolide (**1**).

Because of its ready availability from ergosterol (**51**) without side chain reconstruction, 24-epibrassinolide (**13**) is the most important candidate for practical application, although its bioactivity in some bioassays is lower than those of brassinolide (**1**). Based on earlier studies of other laboratories, an improved synthesis of **13** was published by McMORRIS and PATIL (*122*) (Scheme 7). In this sequence starting compound **51** was transformed via solvolysis of its mesylate to the corresponding 3α,5-cyclo-6β-hydroxy compound followed by oxidation with CrO_3-pyridine to yield the cyclo-6-ketone **52**. Selective reduction of the Δ^7-double bond with lithium in liquid ammonia and subsequent isomerization by heating with pyridinium hydrochloride and LiBr in dimethylacetamide afforded (22*E*,24*R*)-5α-ergosta-2,22-dien-6-one (**53**) as suitable intermediate for the subsequent Sharpless dihydroxylation procedure. Thus, reaction of **53** with OsO_4 in presence of dihydroquinidine 4-chlorobenzoate as chiral ligand improved the yield of the desired (22*R*,23*R*) diastereomer 24-epicastasterone (**14**), which was obtained in 80% yield. Direct Baeyer-Villiger oxidation with trifluoroperoxy acetic acid led to 24-epibrassinolide (**13**). In such a manner target

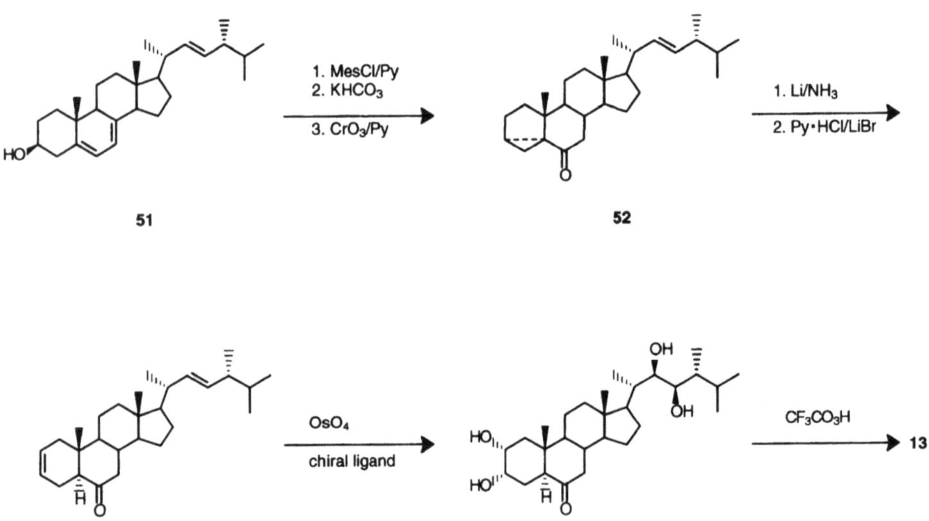

51 52 53 14

Scheme 7. Synthesis of 24-epibrassinolide (**13**) by McMORRIS and PATIL (1993)

compound **13** is available from ergosterol (**51**) in only 7 steps with an overall yield of 26% which predestines this reaction sequence also for larger scale preparation.

7. Biosynthesis

7.1. Biosynthesis in *Catharanthus roseus*

The biosynthesis of brassinosteroids has been studied mainly during the last six years. The results have been summarized in a number of recent reviews (*8–10*, *123*, *124*). Detailed investigations on the biosynthesis of brassinolide (**1**) have been carried out using both non-transformed and crown-gall cell cultures of *Catharanthus roseus* which were shown to produce suitable amounts of brassinosteroids (*32*). It seems likely that brassinosteroids are biosynthesized from phytosterol precursors bearing the corresponding type of side chain. Consequently, a number of parallel pathways may operate even in one species. Except for the 24*S*-methyl side chain bearing brassinosteroids, represented *e.g.* by castasterone (**3**) and brassinolide (**1**), this has not yet been experimentally demonstrated. While the carbon skeleton of **1** is already preformed in campesterol (**54**), the first biosynthetic intermediate, campestanol (**55**), is formed by reduction of the $\Delta^{5,6}$ double bond to afford the trans-fused A/B-ring system of the 5α-cholestane skeleton which is a common characteristical structural feature of all brassinosteroids. Campesterol (**54**), accounting for 50% of the total phytosterol contents of these cultured cells (*125*), was biosynthetically prepared in, isolated from and re-administered to *C. roseus* cell cultures in labelled form (^{13}C and ^{14}C, respectively). All further steps of the biosynthetic sequence finally affording **1** either represent introduction to or modifications of oxygen functionalities on the scaffold. Hydroxylations take place both at the ring system (C-6 and C-2) and at the side chain (C-22 and C-23). Further oxidation of the hydroxyl at C-6 to form a keto group and inversion of configuration of the hydroxyl group at C-3 take place at the ring system. However, the order of these hydroxylation reactions may be different even in the same species. In the so-called early C-6-oxidation pathway hydroxylation first takes place at the B-ring initiating the sequence from 6α-hydroxycampestanol (**56**) to castasterone (**3**) (Scheme 8).

This pathway was established by stepwise conversion of labelled intermediates. For example, after feeding of [^{14}C]- and/or [^{13}C] campesterol (**54**) to cell cultures of *C. roseus*, campestanol (**55**), 6α-hydroxycampestanol (**56**) and 6-oxocampestanol (**57**) were found (*125*).

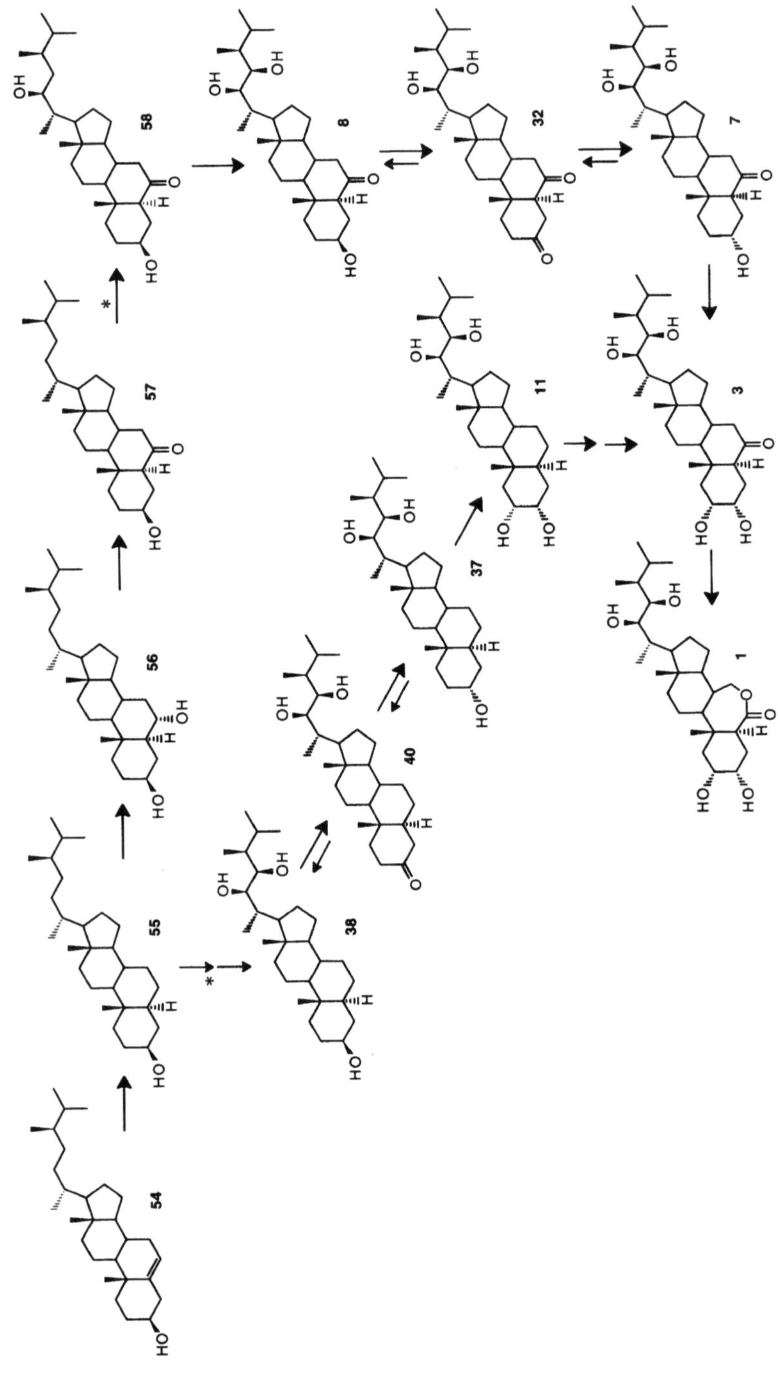

Scheme 8. Brassinosteroid biosynthesis *via* the early and late C-6 oxidation pathway (steps marked by an asterisk are putative)

References, pp. 36–46

Application of [^{2}H]cathasterone (**58**) afforded teasterone (**8**) (*126*) while after administration of [^{2}H]-**8** typhasterol (**7**) and after administration of [^{2}H]-**7** castasterone (**3**) and brassinolide (**1**) were detected by GC-MS (*127*). Finally, in a key step of brassinosteroid biosynthesis [^{3}H]- and/or [^{2}H]-**3** was converted to **1** by a Baeyer-Villiger type oxidation (*128, 129*). Only the conversion of **57** to **58** remains tentative. This might be due to the very small endogeneous pool size (2–4 ng g^{-1} dry mass) of **58** in *C. roseus* cell cultures which is significantly below the level of the most other biosynthetic intermediates (*126*). Otherwise the sequence from **57** *via* 23α-hydroxycampestanol to **8** cannot be ruled out as an alternative partial pathway. The gradual increase of the bioactivity (*130*) from **57** to **1** supported the pathway shown in Scheme 8.

In the late C-6-oxidation pathway, which has been established as an alternative biosynthetic sequence also operating in *C. roseus*, side chain hydroxylation at C-22 and C-23 as well as hydroxylation at C-2 occur prior to introduction of the keto group at C-6 (*34*, *131*). This was again demonstrated by sequential feeding experiments: 6-Deoxo-teasterone (**38**) was converted to 6-deoxo-typhasterol (**37**). 3-Dehydro-6-deoxo-teasterone (**40**) and **37** were converted to 6-deoxo-castasterone (**11**), and, for its part, the latter was transformed to **3**. Since **38** and **37** were not converted to teasterone (**8**) and typhasterol (**7**), respectively, it is most likely that in the late C-6 oxidation pathway only **11** undergoes oxidation at C-6 to give **3** (*132*). Just like in the early C-6-oxidation variant in which **57** is the tentative precursor of **8** the actual sequence of the side chain hydroxylation in the late C-6 oxidation between **55** and **38** remains to be examined.

7.2. Biosynthesis in Other Species

The co-occurrence of teasterone (**8**), typhasterol (**7**), castasterone (**3**) and brassinolide (**1**) in several plant species (see Table 1) indicates that the early C-6 oxidation pathway seems to be generally operating in the plant kingdom. The late C-6 oxidation pathway seems to be also operative as an alternative to the early C-6 oxidation variant in species other than *C. roseus*. This was supported by the occurrence of several 6-deoxo type brassinosteroids especially 6-deoxocastasterone (**11**) which recently have been found (see Table 1). However, lactonization of the B-ring (Baeyer-Villiger reaction) which is operating after the branched part of the brassinosteroid biosynthesis could not be demonstrated in species other than *C. roseus*. Thus, intact seedlings *of Nicotiana tabacum*

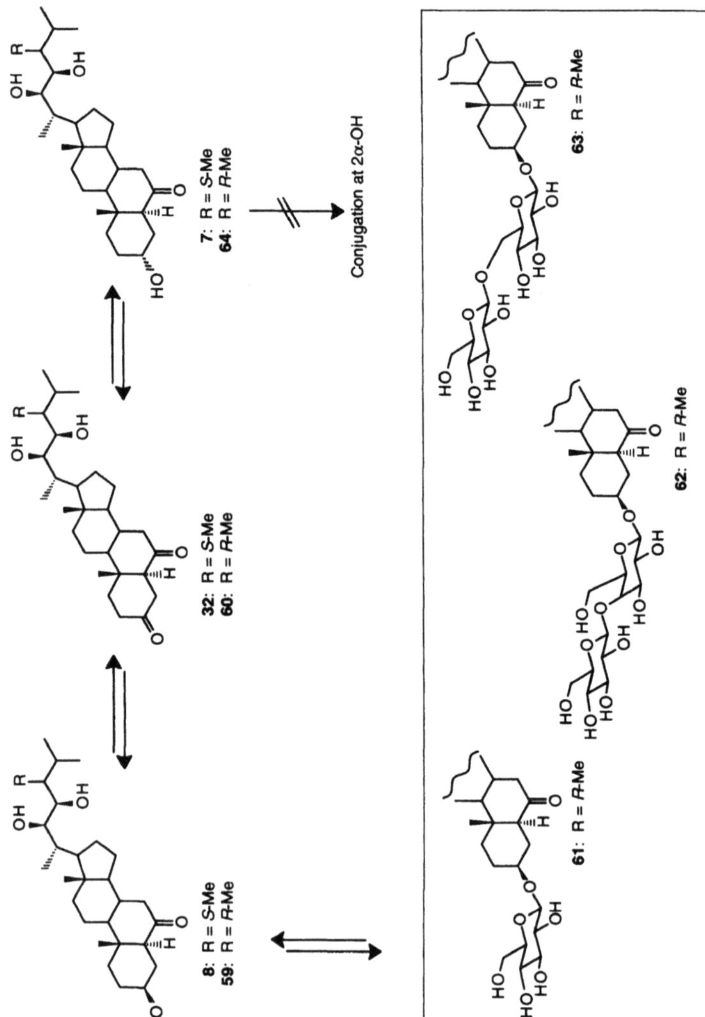

Scheme 9. Reversible reactions involved in brassinosteroid biosynthesis

and *Oryza sativa* converted **8** to **7** and **7** to **3**, but attempts to transform **3** further to **1** failed in these species (*132*).

The reversibility of the epimerization of the 3-hydroxyl group has been described several times (*34*, *132–134*). 24-Epiteasterone (**59**), located in the biosynthetic sequence before the inversion of 3β-OH to the 3α-OH configuration, and 3-dehydro-24-epiteasterone (**60**), was found to be transformed to the glucoside **61** and the two diglycosides **62** and **63** (*134*) (Scheme 9). Glycosidation which in general is a reversible conjugation reaction occurred exclusively at the 3β position as indicated by NMR studies. It was assumed that both 3β-configuration and inversion of configuration at the 3-OH might be involved in regulation of the pool size of the biosynthetic intermediates, 24-epiteasterone (**59**) and 24-epityphasterol (**64**).

8. Metabolism of Brassinosteroids

8.1. Metabolism in Plants

Currently more than 30 brassinosteroid metabolites and conjugates are known, most of them being new compounds not yet found in plants (for reviews see *7*, *135*). Many of these metabolites have to be considered as deactivation or even degradation products. However, some of them might be involved in the mode of action of brassinosteroids. This has been discussed in particular for 25-hydroxy-24-epibrassinolide (**66**) (*136*).

A number of metabolic investigations have been undertaken in seedlings of various plant species which occassionally afforded defined metabolites. In seedlings of *Vigna radiata*, for example, brassinolide (**1**) was converted to its 23-O-β-D-glucoside (**65**) (Scheme 10) (*137*) whereas 3-epicastasterone (**5**) was found as a metabolite of castasterone (**3**) in *Catharanthus roseus*, *Nicotiana tabacum* and *Oryza sativa* (*132*).

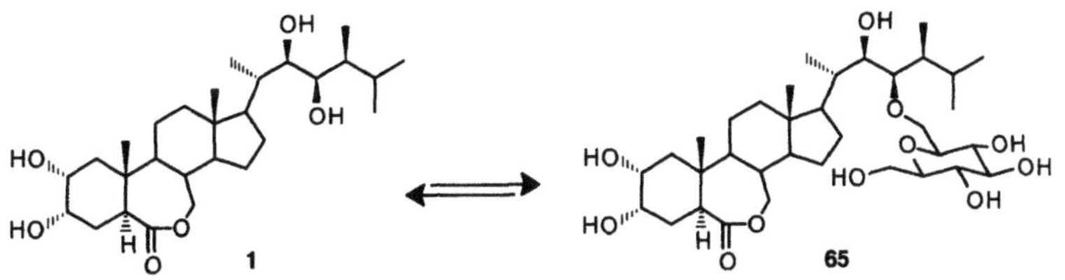

Scheme 10. Glucosidation of brassinolide in *Vigna radiata* seedlings

8.2. Metabolism in Cell Cultures of *Lycopersicon esculentum*

More detailed studies were carried out using cell suspension cultures of *Lycopersicon esculentum*. Side chain hydroxylation at C-25 and C-26, respectively, followed by glucosidation of the new hydroxyl groups are the predominant reactions of 24-epicastasterone (**14**) and 24-epibrassinolide (**13**) in *L. esculentum* cell cultures (*138*, *139*). Only the glucosides **70–73** of 25- and 26-hydroxy-24-epibrassinolide (**66**, **68**) and 25- and 26-hydroxy-24-epicastasterone (**67**, **69**), respectively, were detected as metabolites in cell cultures, from which the aglycones **66–69** were

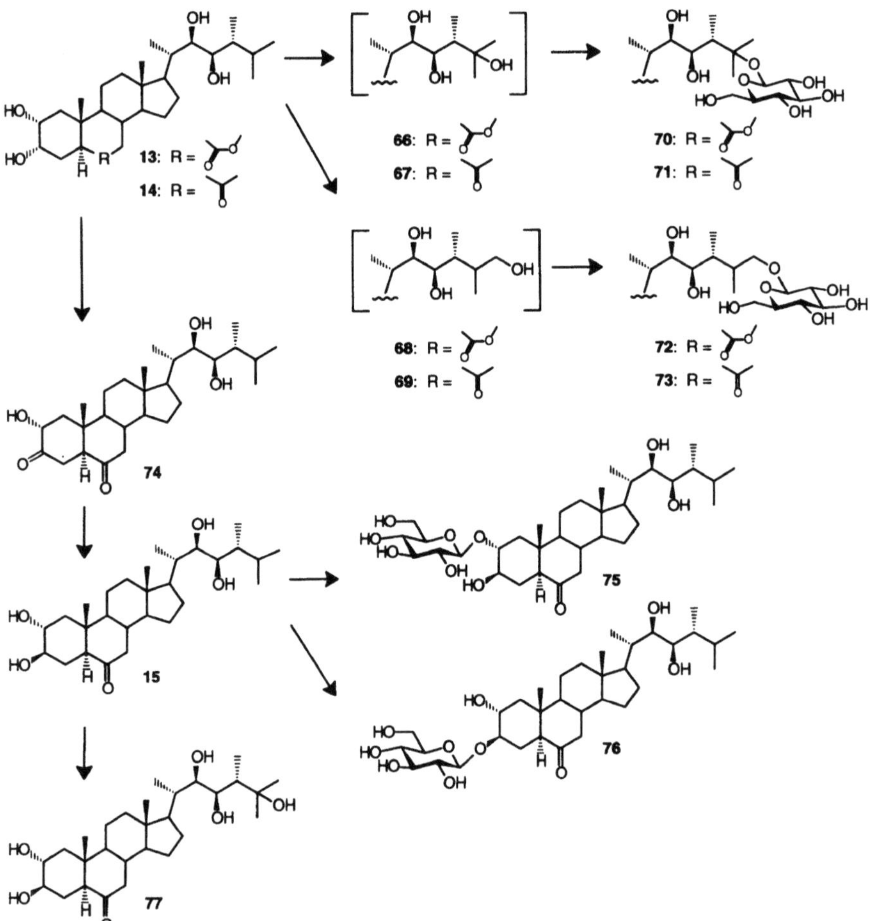

Scheme 11. Metabolism of 24-epicastasterone and 24-epibrassinolide in *Lycopersicon esculentum* cell cultures

obtained by enzymatic hydrolysis. Experiments to characterize the 24-epibrassinolide-C-25 and -C-26 hydroxylase, respectively, by means of cytochrome P450 inhibitors demonstrated that the C-25 hydroxylase is a typical cytochrome P450 enzyme but the C-26 hydroxylase is not P450 dependent. Both proteins were shown to be inducible only by the substrates 24-epibrassinolide (**13**) and brassinolide (**1**) (*140*) and therefore are considered as highly specific enzymes. In *L. esculentum* the 3α-hydroxyl group of 24-epicastasterone (**14**) has been converted to the 3β-hydroxyl in **15** *via* the 3,6-diketone **74**. Subsequently, glucosidation at 2α-hydroxyl and 3β-hydroxyl, respectively, to give compounds **75** and **76**, and hydroxylation at C-25 of **15** to form compound **77** were found (*139*). These metabolites casually fit well into the metabolic pathways shown in Scheme 11.

8.3. Metabolism in Cell Cultures of *Ornithopus sativus*

The metabolism of 24-epicastasterone (**14**) and 24-epibrassinolide (**13**) has been also examined in cell cultures of *Ornithopus sativus* (Fabaceae) (*141*). Isolation of 3,24-diepibrassinolide (**78**) just 24 h after application of [^{3}H]-**13** indicated that epimerization of the 3-hydroxyl group was the initial metabolic step in that species. The corresponding metabolite of the 6-oxo series, 3,24-diepicastasterone (**15**), known as a native brassinosteroid from seeds of *Phaseolus vulgaris* (*69*), was found after application of 24-epicastasterone (**14**). Both 3,24-diepibrassinosteroids **15** and **78** represent branching points in the metabolic pathway shown in Scheme 12. On the one hand, they underwent acyl conjugation with lauric-, myristic-, and palmitic acid (*142*) to give compounds **79**–**81**, and **82**–**84**, respectively. Like glycosidation in tomato cell cultures (*134*, *139*), esterification with fatty acid requires equatorial configuration of the 3-hydroxyl group of the precursor. On the other hand, **15** and **78** were hydroxylated at C-20 to form the pentahydroxylated compounds 20*R*-hydroxy-3,24-diepicastasterone (**85**) and 20*R*-hydroxy-3,24-diepibrassinolide (**86**) which represent the immediate precursors of pregnane type compounds 2α,3β-dihydroxy-B-homo-6a-oxa-5α-pregnane-6,20-dione (**88**) and 2α,3β-dihydroxy-5α-pregnane-6,20-dione (**87**), respectively (*141*, *143*). The mechanism of that side chain cleavage seems to parallel that found for 20-hydroxyecdysones (*144*, *145*). The 6-oxo functionality in compound **87** was further reduced resulting in formation of 2α,3β,6β-trihydroxy-5α-pregnane-20-one (**89**). Hydroxylation of 3,24-diepibrassinolide (**78**) at C-25 to give 25-hydroxy-3,24-diepibrassinolide (**90**) completed the metabolic spectrum observed in *O. sativus* cell

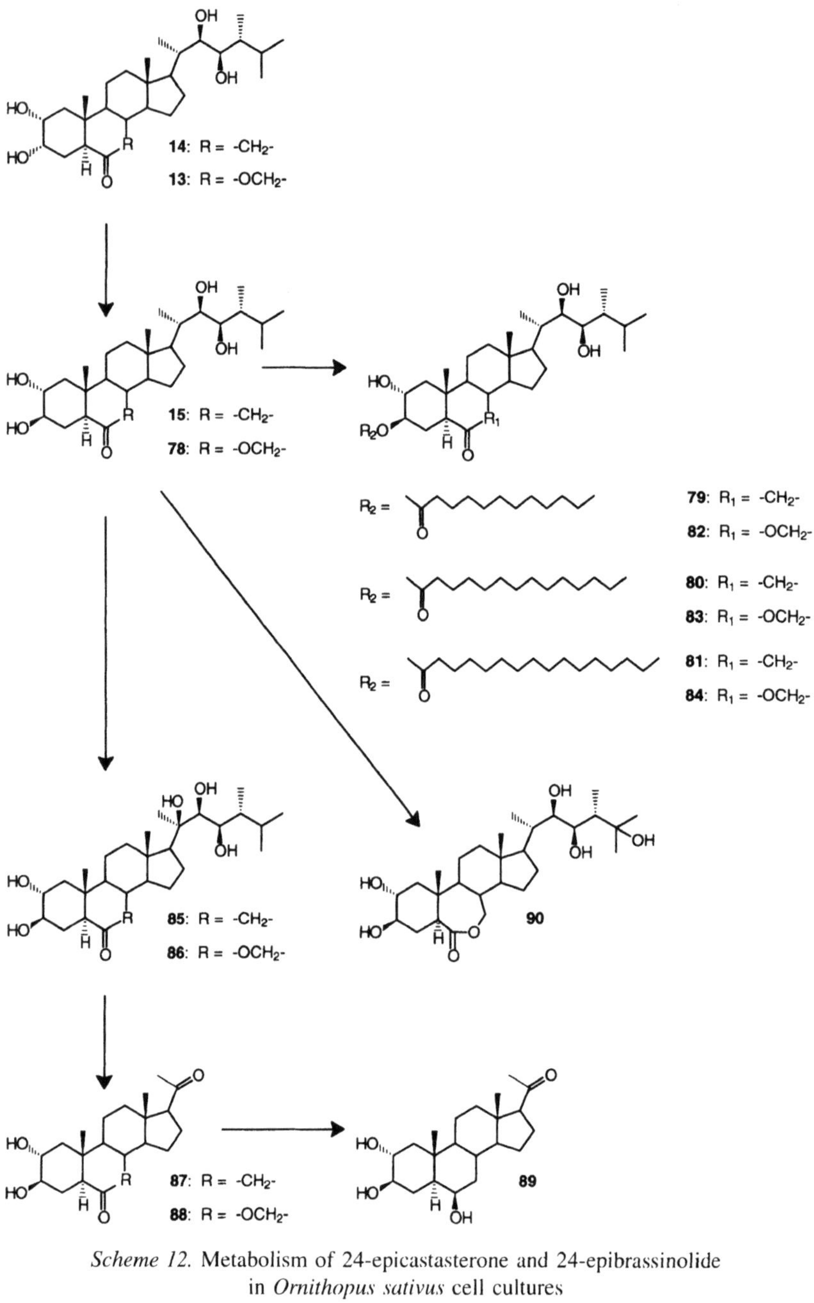

Scheme 12. Metabolism of 24-epicastasterone and 24-epibrassinolide in *Ornithopus sativus* cell cultures

cultures. Conversion of **14** to **13** could not be detected in cell cultures of either *L. esculentum* or *O. sativus*.

9. Physiological Action

The physiological action of plant growth promoting brassinosteroids has been a field of intensive studies with many plant systems and whole plants indicating a broad spectrum of biological activity. Thus, at very low concentrations ($10^{-7}-10^{-12}$ M) biologically active brassinosteroids enhance the growth of young plant tissue and stimulate metabolic, differentiation and growth processes. Among others the observed multiple properties involve effects on cell division, elongation and cell membranes, activation of photosynthesis and specific enzymes as well as interaction with other phytohormones and external signals, *e.g.* light, temperature and gravitropism. A favourable influence of exogenously applied brassinosteroids upon plant stress like drought, temperature, pathogenic infection, herbicidal injury and salinity was also established. Effects on fungi and insects have been reported too. A detailed treatment of these physiological aspects is the topic of many reviews (*2–13*, *146–149*).

The bioactivity of brassinosteroids first of all depends on their specific structure and the applied test system. Previous structure-activity studies (*4*, *150–153*) revealed the following essentials: For high activity a vicinal 22*R*,23*R*-diol function in the side chain moiety is necessary and is present in all hitherto isolated naturally occurring compounds (Fig. 1). Alkylation at C-24 increases the activity in the sequence $CH_3 > C_2H_5 >$ H. In compounds with *trans*-fused A/B rings an 6a-oxalactone or 6-oxo structure is important, with the lactone types exhibiting greater activity. As regards the hydroxylation pattern of ring A a vicinal 2α,3α-diol struc-ture exhibits the greatest potency. For evaluation of structure-activity relationships and putative receptor interactions the first molecular modeling studies of brassinosteroids have been published (*154–157*).

Because of their high biological activity at very low concentrations, non-toxicity and their native character brassinosteroids represent promising candidates for further development as ecologically soft bioregulators and stress modulators. Therefore, intensive efforts have been made towards the practical application of brassinosteroids in agriculture and horticulture, with the aim to increase the crop yield and to use their antistress effects for stimulation of physiological processes in plants under unfavorable conditions (*4*, *9*, *158–161*). Experiments with various crop plants in field trials have progressed especially in Japan, China and some countries of the former USSR and are summarized in (*13*, *162*). Because of its high activity and easy availability most of these studies were carried out with 24-epibrassinolide (**13**). However, some promising structural analogues of brassinosteroids with 22,23-epoxy function which showing longer-lasting activity under field conditions, are now also being tested officially for practical application (*163*, *162*).

10. Molecular Mode of Action

Despite extended biological studies (see Chapter 9) the physiological role of brassinosteroids remained unsolved for many years. This has been changed mainly by studies on dwarf mutants defective in biosynthesis or perception of brassinosteroids thus providing evidence for the essential role of brassinosteroids in plant growth and development (for reviews see *164–167*). A number of mutants of *Arabidopsis thaliana* (*det2*; *dwf4*; *cpd*, allelic with *cbb3*) could be rescued to the wild type phenotype by treatment with exogenously applied brassinosteroids but not by use of other phytohormones and were therefore considered to be deficient in one of the brassinosteroid biosynthetic enzymes. Molecular analysis of genomic mutations revealed the sites of biosynthetic lesions. The *DET2* protein was shown to share about 40% identity with mammalian steroid 5α-reductases (*168*). Thus, it was deduced that the *DET2* gene might be involved in the reduction of campesterol (**54**). In fact it was demonstrated that the campestanol (**55**) content of *det2* mutants is significantly below the wild type level (*169*). More detailed biochemical investigations (*170*) located the gene defect more precisely in the 5α-reduction of (24*R*)-24-methylcholest-4-en-3-one (**91**), which is supposed to be further modified to form **55** via (24*R*)-24-methyl-5α-cholestan-3-one (**92**) (Scheme 13).

Since many steps of the brassinosteroid biosynthesis involve hydroxylations (see Chapter 7) it is likely that mutations occur in gene loci encoding monooxygenases. The *CPD* gene of the *cpd* mutant (*171*) which is allelic with the *cbb3* mutant (*172*) has been cloned and found to encode a protein (CYP90A1) showing high sequence identity with conserved domains of microsomal P450 monooxygenases and some homology with specific domains of steroid hydroxylases (*171*). Together with the finding that brassinosteroids and C-23 hydroxylated biosynthetic intermediates are able to rescue the dwarf mutants to the wildtype phenotype but that more early precursors failed, it was concluded that the *CPD* protein must be involved in the conversion of cathasterone (**58**) to

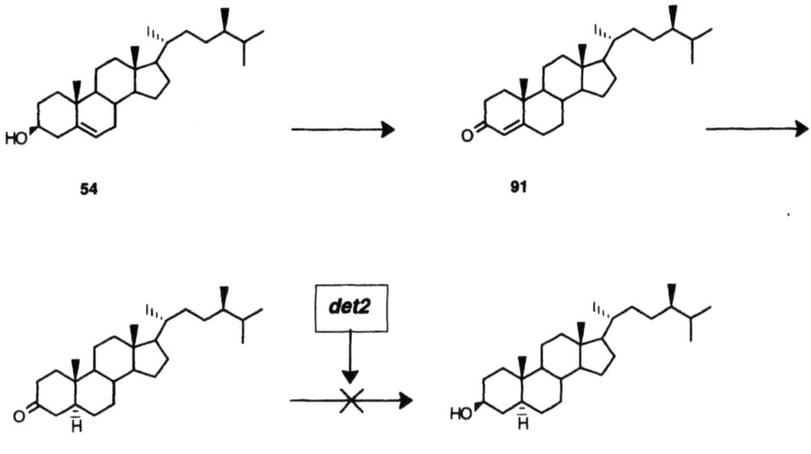

Scheme 13. Proposed early sequence of brassinosteroid biosynthesis

teasterone (**8**). Probably CYP90A1 accepts **58** as well as its 6-deoxo analog as a substrate since otherwise the alternative pathway should be used by the plant cells to synthesize active brassinosteroids (*165*).

Studies on the regulation of brassinosteroid biosynthesis in *Arabidopsis* showed that the transcription of the *CPD* gene is specifically down-regulated by brassinolide (**1**) but not by other plant hormones (*173*). Thus, the control of *CPD* expression by brassinosteroids seems to be analogous to feed-back regulation of steroidogenic cytochrome P450 genes by oxysterols in animals.

Moreover, *DWF4* was cloned and identified by its characteristic signature sequences as a cytochrome P450 (*174*). The *DWF4* gene displays significant homology with CPD and therefore was grouped into the same P450 family and designated CYP90B1. Rescue of *dwf4* by 22α-hydroxylated steroids suggested that hydroxylation of the C-22 position is affected in this mutant, both in early and late C-6 oxidation biosynthetic pathways. It should be mentioned, however, that the results obtained with *Arabidopsis* mutants are plausible only by assuming that the biosynthetic pathways parallel those of *C. roseus*. Several further mutants which probably are brassinosteroid deficient have been found in *Arabidopsis thaliana* (*dim*, allelic with *cbb1*) (*172*, *175*) and *Pisum sativum* (*lkb*) (*176*). However, the site of the lesions in the biosynthetic pathway has not yet been exactly defined.

Brassinosteroid-insensitive mutants which have also been identified from *Arabidopsis* (*bri1*, allelic with *cbb2*) (*172*, *177*, *178*) and *Pisum sativum* (*lka*) (*176*) do not respond to exogeneous application of brassinosteroids with phenotype normalization. These mutations are considered to be affected either in the hormone receptor or the signal transduction chain. However, since it has not been finally proved that brassinolide (**1**) is the actual active hormone, it might also be possible that these mutations block a hypothetical final biosynthetic step.

In conclusion, the mutant work and other molecular genetic studies revealed crucial regulatory functions of brassinosteroids. In particular, brassinosteroids control cell elongation (*179*) probably through lowered expression of *TCH4* genes which encode a xyloglucan endotransglycosylase (XET) (*180*), and photomorphogenesis (*171*).

11. Conclusions

Since the discovery of the first member of this series of compounds, brassinosteroid research has been a field of tremendous research activity, concentrating up to about 1990 mainly on phytochemical analysis, chemical synthesis and biological aspects. Subsequently studies on biosynthesis and metabolism of brassinosteroids became an increasing focus of attention, utilizing mainly plant cell culture systems and setting up the basis for more recent molecular genetic investigations. Carried out especially with brassinosteroid-deficient mutants of *Arabidopsis thaliana*, these studies have caused a breakthrough towards a better understanding of the molecular action of brassinosteroids and have established their essential function for plant growth and development. Thus, as result of comprehensive interdisciplinary cooperation, nowadays brassinosteroids can be regarded as a new class of phytohormones with interesting future perspectives for use of these biological highly active plant growth regulators.

References

1. GROVE, M.D., G.F. SPENCER, W.K. ROWEDDER, N. MANDAVA, J.F. WORLEY, J.D. WARTHEN, Jr., G.L. STEFFENS, J.L. FLIPPEN-ANDERSON, and J.K. COOK, Jr.: Brassinolide, a Plant Growth-Promoting Steroid Isolated from *Brassica napus* Pollen. Nature, **281**, 216 (1979).

2. YOKOTA, T., and N. TAKAHASHI: Chemistry, Physiology and Agriculture Application of Brassinolide and Related Steroids. In: Plant Growth Substances (M. Bopp, ed.), p. 129. Berlin, Heidelberg, New York: Springer-Verlag. 1986.

3. Adam, G., and V. Marquardt: Brassinosteroids. Phytochemistry, **25**, 1787 (1986).

4. Mandava, N.B.: Plant Growth-Promoting Brassinosteroids. Annu. Rev. Plant Physiol. Plant Mol. Biol., **39**, 23 (1988).

5. Marquardt, V., and G. Adam: Recent Advances in Brassinosteroid Research. In: Chemistry of Plant Protection, 7 (W. Ebing, H. Börner, D. Martin, V. Sjut, H.-J. Stan, and J. Stetter, eds.), p. 103. Berlin, Heidelberg: Springer-Verlag. 1991.

6. Sakurai, A., and S. Fujioka: The Current Status of Physiology and Biochemistry of Brassinosteroids. Plant Growth Regul., **13**, 147 (1993).

7. Adam, G., A. Porzel, J. Schmidt, B. Schneider, and B. Voigt: New Developments in Brassinosteroid Research. In: Studies in Natural Products Chemistry (Atta-ur-Rahman, ed.), p. 495. Amsterdam: Elsevier Science. 1996.

8. Yokota, T.: Structure, Biosynthesis and Function of Brassinosteroids. Trends Plant Sci., **2**, 137 (1997).

9. Sasse, J.M.: Recent Progress in Brassinosteroid Research. Physiol. Plant., **100**, 696 (1997).

10. Fujioka, S., and A. Sakurai: Brassinosteroids. Nat. Prod. Rep., **14**, 1 (1997).

11. Clouse, S.D., and J.M. Sasse: Brassinosteroids: Essential Regulators of Plant Growth and Development. Annu. Rev. Plant Physiol. Plant Mol. Biol., **49**, 427 (1998).

12. Cutler, H.G., T. Yokota, and G. Adam (eds.): Brassinosteroids – Chemistry, Bioactivity, and Applications, Washington, DC: American Chemical Society. 1991.

13. Khripach, V.A., V.N. Zhabinskii, and A.E. de Groot: Brassinosteroids – A New Class of Plant Hormones. San Diego: Academic Press. 1999.

14. Sakurai, A., T. Yokota, and S. Clouse (eds.): Brassinosteroids, Steroidal Plant Hormones. Tokyo: Springer-Verlag. 1999.

15. Plattner, R.D., S.L. Taylor, and M.D. Grove: Detection of Brassinolide and Castasterone in *Alnus glutinosa* (European Alder) by Mass Spectrometry/Mass Spectrometry. J. Nat. Prod., **49**, 540 (1986).

16. Schmidt, J., B. Voigt, and G. Adam: 2-Deoxybrassinolide – A Naturally Occurring Brassinosteroid from *Apium graveolens*. Phytochemistry, **40**, 1041 (1995).

17. Fujioka, S., Y.-H. Choi, S. Takatsuto, T. Yokota, J. Li, J. Chory, and A. Sakurai: Identification of Castasterone, 6-Deoxocastasterone, Typhasterol and 6-Deoxotyphasterol from Shoots of *Arabidopsis thaliana*. Plant Cell Physiol., **37**, 1201 (1996).

18. Schmidt, J., T. Altmann, and G. Adam: Brassinosteroids from Seeds of *Arabidopsis thaliana*. Phytochemistry, **45**, 1325 (1997).

19. Fujioka, S., T. Noguchi, T. Yokota, S. Takatsuto, and S. Yoshida: Brassinosteroids in *Arabidopsis thaliana*. Phytochemistry, **48**, 595 (1998).

20. Takatsuto, S.: Brassinosteroids: Distribution in Plants, Bioassay and Microanalysis by Gas Chromatography-Mass Spectrometry. J. Chromatogr. A, **658**, 3 (1994).

21. Schmidt, J., C. Kuhnt, and G. Adam: Brassinosteroids and Sterols from *Beta vulgaris*. Phytochemistry, **36**, 175 (1994).

22. Ikekawa, N., S. Takatsuto, T. Kitsuwa, H. Saito, T. Morishita, and H. Abe: Analysis of Natural Brassinosteroids by Gas Chromatography and Gas Chromatography-Mass Spectrometry. J. Chromatogr., **290**, 289 (1984).

23. Abe, H., T. Morishita, M. Uchiyama, S. Marumo, K. Munakata, S. Takatsuto, and N. Ikekawa: Identification of Brassinolide-Like Substances in Chinese Cabbage. Agric. Biol. Chem., **46**, 2609 (1982).

24. Ikekawa, N., S. Takatsuto, S. Marumo, H. Abe, T. Morishita, M. Uchiyama, M. Ikeda, T. Sassa, and T. Kitsuwa: Identification of Brassinolide and its 6-Oxo Analog in Plant Kingdom by Selected Ion Monitoring. Proc. Japan Acad. Ser. B, **59**, 9 (1983).

25. ABE, H., T. MORISHITA, M. UCHIYAMA, S. TAKATSUTO, N. IKEKAWA, M. IKEDA, T. SASSA, T. KITSUWA, and S. MARUMO: Occurrence of Three New Brassinosteroids: Brassinone, (24*S*)-24-Ethylbrassinone and 28-Norbrassinone, in Higher Plants. Experientia, **39**, 351 (1983).
26. TAKATSUTO, S., H. ABE, T. YOKOTA, K. SHIMADA, and K. GAMOH: Identification of Castasterone and Teasterone in Seeds of *Cannabis sativa* L. Nihon Yukagakkaishi, **45**, 871 (1996).
27. GAMOH, K., H. ABE, K. SHIMADA, and S. TAKATSUTO: Liquid Chromatography/Mass Spectrometry with Atmospheric Pressure Chemical Ionization of Free Brassinosteroids. Rap. Commun. Mass Spectrom., **10**, 903 (1996).
28. PARK, K.-H., J.-D. PARK, K.-H. HYUN, M. NAKAYAMA, and T. YOKOTA: Brassinosteroids and Monoglycerides in Immature Seeds of *Cassia tora* as the Active Principles in the Rice Lamina Inclination Bioassay. Biosci. Biotechnol. Biochem., **58**, 1343 (1994).
29. YOKOTA, T., M. ARIMA, and N. TAKAHASHI: Castasterone, a New Phytosterol with Plant-Hormone Potency, from Chestnut Insect Gall. Tetrahedron Letters, **23**, 1275 (1982).
30. ARIMA, M., T. YOKOTA, and N. TAKAHASHI: Identification and Quantification of Brassinolide-Related Steroids in the Insect Gall and Healthy Tissues of the Chestnut Plant. Phytochemistry, **23**, 1587 (1984).
31. IKEDA, M., S. TAKATSUTO, T. SASSA, N. IKEKAWA, and M. NUKINA: Identification of Brassinolide and its Analogues in Chestnut Gall Tissue. Agric. Biol. Chem., **47**, 655 (1983).
32. PARK, K.-H., H. SAIMOTO, S. NAKAGAWA, A. SAKURAI, T. YOKOTA, N. TAKAHASHI, and K. SYONO: Occurrence of Brassinolide and Castasterone in Crown Gall Cells of *Catharanthus roseus*. Agric. Biol. Chem., **53**, 805 (1989).
33. CHOI, Y.-H., T. INOUE, S. FUJIOKA, H. SAIMOTO, and A. SAKURAI: Identification of Brassinosteroid-Like Substances in Plant Cell Cultures. Biosci. Biotechnol. Biochem., **57**, 860 (1993).
34. CHOI, Y.-H., S. FUJIOKA, T. NOMURA, A. HARADA, T. YOKOTA, S. TAKATSUTO, and A. SAKURAI: An Alternative Brassinolide Biosynthetic Pathway via Late C-6 Oxidation. Phytochemistry, **44**, 609 (1997).
35. MOTEGI, C., S. TAKATSUTO, and K. GAMOH: Identification of Brassinolide and Castasterone in the Pollen of Orange (*Citrus sinensis* Osbeck). J. Chromatogr. A, **658**, 27 (1994).
36. ABE, H: Rice-Lamina Inclination, Endogenous Levels in Plant Tissues and Accumulation During Pollen Development of Brassinosteroids. In: ref. 12, p. 200.
37. YOKOTA, T., K. HIGUCHI, N. TAKAHASHI, Y. KAMURO, T. WATANABE, and S. TAKATSUTO: Identification of Brassinosteroids with Epimeric Substituents and/or the 23-Oxo Group in Pollen and Anthers of *Japanese Cedar*. Biosci. Biotechnol. Biochem., **62**, 526 (1998).
38. SCHMIDT, J., and G. ADAM, unpublished results.
39. GRIFFITHS, P.G., J.M. SASSE, T. YOKOTA, and D.W. CAMERON: 6-Deoxotyphasterol and 3-Dehydro-6-deoxoteasterone, Possible Precursors to Brassinosteroids in the Pollen of *Cupressus arizonica*. Biosci. Biotechnol. Biochem., **59**, 956 (1995).
40. SCHMIDT, J., A. PORZEL, and G. ADAM: Brassinosteroids and a Pregnane Glucoside from *Daucus carota*. Phytochem. Anal., **9**, 14 (1998).
41. YOKOTA, T.: Study on Bioorganic Chemistry of Natural Brassinosteroids. Chem. Regul. Plants, Tokyo (Shokobutsu no Kagaku Chosetsu), **22**, 10 (1987).

42. ABE, H., C. HONJO, Y. KYOKAWA, S. ASAKAWA, M. NATSUME, and M. NARUSHIMA: 3-Oxoteasterone and the Epimerization of Teasterone: Identification in Lily Anthers and *Distylium racemosum* Leaves and its Biotransformation into Typhasterol. Biosci. Biotechnol. Biochem., **58**, 986 (1994).
43. YOKOTA, T., J. BABA, and N. TAKAHASHI: A new Steroidal Lactone with Plant Growth-Regulatory Activity from *Dolichos lablab* Seed. Tetrahedron Letters, **23**, 4965 (1982).
44. BABA, J., T. YOKOTA, and N. TAKAHASHI: Brassinolide-Related New Bioactive Steroids from *Dolichos lablab* Seed. Agric. Biol. Chem., **47**, 659 (1983).
45. YOKOTA, T., J. BABA, and N. TAKAHASHI: Brassinolide-related Bioactive Sterols in *Dolichos lablab*: Brassinolide, Castasterone and a New Analog, Homodolicholide. Agric. Biol. Chem., **47**, 1409 (1983).
46. YOKOTA, T., J. BABA, S. KOBA, and N. TAKAHASHI: Purification and Separation of Eight Steroidal Plant Growth Regulators from *Dolichos lablab* Seed. Agric. Biol. Chem., **48**, 2529 (1984).
47. TAKATSUTO, S., H. ABE, and K. GAMOH: Evidence for Brassinosteroids in Strobilus of *Equisetum arvense* L. Agric. Biol. Chem., **54**, 1057 (1990).
48. YASUTA, E., T. TERAHATA, M. NAKAYAMA, H. ABE, S. TAKATSUTO, and T. YOKOTA: Free and Conjugated Brassinosteroids in the Pollen and Anthers of *Erythronium japonicum* Decne. Biosci. Biotechnol. Biochem., **59**, 2156 (1995).
49. TAKATSUTO, S., K. OMOTE, K. GAMOH, and M. ISHIBASHI: Identification of Brassinolide and Castasterone in Buckwheat (*Fagopyrum esculentum* Moench) Pollen. Agric. Biol. Chem., **54**, 757 (1990).
50. TAKATSUTO, S., H. ABE, K. SHIMADA, M. NAKAYAMA, and T. YOKOTA: Identification of Teasterone and 4-Desmethylsterols in the Seeds of *Ginkgo biloba* L. Nihon Yukagakkaishi, **45**, 1349 (1996).
51. SCHMIDT, J., F. BÖHME, and G. ADAM: 24-Epibrassinolide from *Gypsophila perfoliata*. Z. Naturforsch., **51c**, 897 (1996).
52. TAKATSUTO, S., T. YOKOTA, K. OMOTE, K. GAMOH, and N. TAKAHASHI: Identification of Brassinolide, Castasterone and Norcastasterone (Brassinone) in Sunflower (*Helianthus annuus* L.) Pollen. Agric. Biol. Chem., **53**, 2177 (1989).
53. YOKOTA, T., S.-K. KIM, Y. FUKUI, N. TAKAHASHI, Y. TAKEUCHI, and T. TAKEMATSU: Brassinosteroids and Sterols from a Green Alga, *Hydrodictyon reticulatum*: Configuration at C-24. Phytochemistry, **26**, 503 (1987).
54. SUZUKI, H., S. FUJIOKA, T. YOKOTA, N. MUROFUSHI, and A. SAKURAI: Identification of Brassinolide, Castasterone, Typhasterol, and Teasterone from the Pollen of *Lilium elegans*. Biosci. Biotechnol. Biochem., **58**, 2075 (1994).
55. ASAKAWA, S., H. ABE, Y. KYOKAWA, S. NAKAMURA, and M. NATSUME: Teasterone 3-myristate: A New Type of Brassinosteroid Derivative in *Lilium longiflorum* Anthers. Biosci. Biotechnol. Biochem., **58**, 219 (1994).
56. ASAKAWA, S., H. ABE, N. NISHIKAWA, M. NATSUME, and M. KOSHIOKA: Purification and Identification of New Acyl-conjugated Teasterones in Lily Pollen. Biosci. Biotechnol. Biochem., **60**, 1416 (1996).
57. TAYLOR, P.E., K. SPUCK, P.M. SMITH, J.M. SASSE, T. YOKOTA, P.G. GRIFFITHS, and D.W. CAMERON: Detection of Brassinosteroids in Pollen of *Lolium perenne* by Immunocytochemistry. Planta, **189**, 91 (1993).
58. YOKOTA, T., T. NOMURA, and M. NAKAYAMA: Identification of Brassinosteroids that Appear to be Derived from Campesterol and Cholesterol in Tomato Shoots. Plant Cell Physiol., **38**, 1291 (1997).

59. FRIEBE, A., A. VOLZ, J. SCHMIDT, B. VOIGT, G. ADAM, and H. SCHNABL: 24-Episecasterone and 24-epi-Castasterone from *Lychnis viscaria* Seeds. In press.
60. SCHMIDT, J., B. SPENGLER, T. YOKOTA, and G. ADAM: The Co-Occurrence of Castasterone and 24-Epicastasterone in Seeds of *Ornithopus sativus*. Phytochemistry, **32**, 1614 (1993).
61. SPENGLER, B., J. SCHMIDT, B. VOIGT, and G. ADAM: 6-Deoxo-28-norcastasterone and 6-Deoxo-24-epicastasterone – Two New Brassinosteroids from *Ornithopus sativus*. Phytochemistry, **40**, 907 (1995).
62. ABE, H., K. NAKAMURA, T. MORISHITA, M. UCHIYAMA, S. TAKATSUTO, and N. IKEKAWA: Endogenous Brassinosteroids of the Rice Plant: Castasterone and Dolichosterone. Agric. Biol. Chem., **48**, 1103 (1984).
63. ABE, H., S. TAKATSUTO, M. NAKAYAMA, and T. YOKOTA: 28-Homotyphasterol, a New Natural Brassinosteroid from Rice (*Oryza sativa* L.) Bran. Biosci. Biotechnol. Biochem., **59**, 176 (1995).
64. PARK, K.-H., J.-D. PARK, K.-H. HYUN, M. NAKAYAMA, and T. YOKOTA: Brassinosteroids and Monoglycerides with Brassinosteroid-Like Activity in Immature Seeds of *Oryza sativa* and *Perilla frutescens* and in Cultured Cells of *Nicotiana tabacum*. Biosci. Biotechnol. Biochem., **58**, 2241 (1994).
65. SHIMADA, K., H. ABE, S. TAKATSUTO, M. NAKAYAMA, and T. YOKOTA: Identification of Castasterone and Teasterone from Seeds of Canary Grass (*Phalaris canariensis*). Rec. Res. Develop. Chem. Pharm. Sci., **1**, 1 (1996).
66. SUZUKI, Y., I. YAMAGUCHI, and N. TAKAHASHI: Identification of Castasterone and Brassinone from Immature Seeds of *Pharbitis purpurea*. Agric. Biol. Chem., **49**, 49 (1985).
67. YOKOTA, T., S.-K. KIM, Y. OGINO, and N. TAKAHASHI: Various Brassinosteroids from *Phaseolus vulgaris* Seeds: Structures and Biological Activity. Proc. 14th Annual Meeting Plant Growth Regulator Society of America, Honolulu, p. 28. 1987.
68. YOKOTA, T., S.-K. KIM, Y. KOSAKA, Y. OGINO, and N. TAKAHASHI: Conjugation of Brassinosteroids. In: Conjugated Plant Hormones – Structure, Metabolism and Function (K. SCHREIBER, H.-R. SCHÜTTE, and G. SEMBDNER, eds), Proc. International Symposium, Berlin. VEB Deutscher Verlag der Wissenschaften. p. 288. 1987.
69. KIM, S.-K.: Natural Occurrences of Brassinosteroids. In: ref. 12, p. 26.
70. YOKOTA, T., M. MORITA, and N. TAKAHASHI: 6-Deoxocastasterone and 6-Deoxodolichosterone: Putative Precursors for Brassinolide-Related Steroids from *Phaseolus vulgaris*. Agric. Biol. Chem., **47**, 2149 (1983).
71. YOKOTA, T., S. KOBA, S.-K. KIM, S. TAKATSUTO, N. IKEKAWA, M. SAKAKIBARA, K. OKADA, K. MORI, and N. TAKAHASHI: Diverse Structural Variations of the Brassinosteroids in *Phaseolus vulgaris* Seed. Agric. Biol. Chem., **51**, 1625 (1987).
72. KIM, S.-K., T. YOKOTA, and N. TAKAHASHI: 25-Methyldolichosterone, a New Brassinosteroid with a Tertiary Butyl Group from Immature Seeds of *Phaseolus vulgaris*. Agric. Biol. Chem., **51**, 2303 (1987).
73. ZAKI, A.K., J. SCHMIDT, F.M. HAMMOUDA, and G. ADAM: Steroidal Constituents from Pollen Grains of *Phoenix dactylifera*. Planta Med. (Suppl.), **59**, A 613 (1993).
74. YOKOTA, T., M. ARIMA, N. TAKAHASHI, and A. CROIZIER: Steroidal Plant Growth Regulators, Castasterone and Typhasterol (2-Deoxycastasterone) from the Shoots of Sitka Spruce (*Picea sitchensis*). Phytochemistry, **24**, 1333 (1985).
75. KIM, S.-K., H. ABE, C.H.A. LITTLE, and R.P. PHARIS: Identification of Two Brassinosteroids from the Cambial Region Scots Pine (*Pinus silverstris*) by Gas Chromatography-Mass Spectrometry, after Detection using a Dwarf Rice Lamina Inclination Bioassay. Plant Physiol., **94**, 1709 (1990).

76. YOKOTA, T., M. ARIMA, N. TAKAHASHI, S. TAKATSUTO, N. IKEKAWA and T. TAKEMATSU: 2-Deoxycastasterone, a new Brassinolide-Related Bioactive Steroid from Pinus Pollen. Agric. Biol. Chem., **47**, 2419 (1983).
77. YOKOTA, T., T. MATSUOKA, T. KOARAI, and M. NAKAYAMA: 2-Deoxybrassinolide, a Brassinosteroid from *Pisum sativum* Seed. Phytochemistry, **42**, 509 (1996).
78. TAKAHITO, N., M. NAKAYAMA, J. B. REID, Y. TAKEUCHI, and T. YOKOTA: Blockage of Brassinosteroid Biosynthesis and Sensitivity Causes Dwarfism in Garden Pea. Plant Physiol., **113**, 31 (1997).
79. SCHMIDT, J., T. YOKOTA, G. ADAM, and N. TAKAHASHI: Castasterone and Brassinolide in *Raphanus sativus* Seeds. Phytochemistry, **30**, 364 (1991).
80. SCHMIDT, J., T. YOKOTA, B. SPENGLER, and G. ADAM: 28-Homoteasterone – a new Naturally Occurring Brassinosteroid from Seeds of *Raphanus sativus*. Phytochemistry, **34**, 391 (1993).
81. SCHMIDT, J., U. HIMMELREICH, and G. ADAM: Brassinosteroids, Sterols and Lup-20(29)-en-2α,3β,28-triol from *Rheum rhabarbarum*. Phytochemistry, **40**, 527 (1995).
82. ABE, H., S. TAKATSUTO, R. OKUDA, and T. YOKOTA: Identification of Castasterone, 6-Deoxocastasterone and Typhasterol in the Pollen of *Robinia pseudo-acacia* L. Biosci. Biotechnol. Biochem., **59**, 309 (1995).
83. SCHMIDT, J., B. SPENGLER, T. YOKOTA, M. NAKAYAMA, S. TAKATSUTO, B. VOIGT, and G. ADAM: Secasterone, the First Naturally Occurring 2,3-Epoxybrassinosteroid from *Secale cereale*. Phytochemistry, **38**, 1095 (1995).
84. MORISHITA, T., H. ABE, M. UCHIYAMA, S. MARUMO, S. TAKATSUTO, and N. IKEKAWA: Evidence for Plant Growth Promoting Brassinosteroids in Leaves of *Thea sinensis*. Phytochemistry, **22**, 1051 (1983).
85. ABE, H., T. MORISHITA, M. UCHIYAMA, S. TAKATSUTO, and N. IKEKAWA: A new Brassinolide-Related Steroid in Leaves of *Thea sinensis*. Agric. Biol. Chem., **48**, 2171 (1984).
86. YOKOTA, T., M. NAKAYAMA, T. WAKISAKI, J. SCHMIDT, and G. ADAM: 3-Dehydroteasterone, a 3,6-Diketobrassinosteroid as a Possible Biosynthetic Intermediate of Brassinolide from Wheat Grain. Biosci. Biotechnol. Biochem. **58**, 1183 (1994).
87. SCHNEIDER, J.A., K. YOSHIHARA, K. NAKANISHI, and N. KATO: Typhasterol (2-Deoxycastasterone): A New Plant Growth Regulator from Cat-Tail Pollen. Tetrahedron Letters, **24**, 3859 (1983).
88. PARK, K.-H., T. YOKOTA, A. SAKURAI, and N. TAKAHASHI: Occurrence of Castasterone, Brassinolide and Methyl 4-Chloroindole-3-acetate in Immature *Vicia faba* Seeds. Agric. Biol. Chem., **51**, 3081 (1987).
89. IKEKAWA, N., F. NISHIYAMA, and Y. FUJIMOTO: Identification of 24-Epibrassinolide in Bee Pollen of the Broad Bean, *Vicia faba* L.. Chem. Pharm. Bull., **36**, 405 (1988).
90. GAMOH, K., N. OMOTE, N. OKAMOTO, and S. TAKATSUTO: High-Performance Liquid Chromatography of Brassinosteroids in Plants with Derivatization Using 9-Phenanthreneboronic Acid. J. Chromatogr., **469**, 424 (1989).
91. SUZUKI, Y., I. YAMAGUCHI, T. YOKOTA, and N. TAKAHASHI: Identification of Castasterone, Typhasterol and Teasterone from the Pollen of *Zea mays*. Agric. Biol. Chem., **50**, 3133 (1986).
92. GAMOH, K., N. OKAMOTO, S. TAKATSUTO, and I. TEIJIMA: Determination of Traces of Natural Brassinosteroids as Dansylaminophenylboronates by Liquid Chromatography with Fluorimetric Detection. Anal. Chim. Acta, **228**, 101 (1990).
93. PARK, K.-H., S.-J. KIM, and T. YOKOTA: Identification of Brassinosteroids in Immature Seeds of *Zea mays* by GC-MS Analysis. Han'guk Nonghwa Hakhoechi, **38**, 179 (1995).

94. WADA, K., S. MARUMO, N. IKEKAWA, M. MORISAKI, and K. MORI: Brassinolide and Homobrassinolide Promotion of Lamina Inclination of Rice Seedlings. Plant Cell Physiol., **22**, 323 (1981).
95. WADA, K., S. MARUMO, H. ABE, T. MORISHITA, K. NAKAMURA, M. UCHIYAMA, and K. MORI: A Rice Lamina Inclination Test – a Micro-Quantitative Bioassay for Brassinosteroids. Agric. Biol. Chem., **48**, 719 (1984).
96. WADA, K., H. KONDO, and S. MARUMO: A Simple Bioassay of Brassinosteroids: A Wheat Leaf-Unrolling Test. Agric. Biol Chem., **49**, 2249 (1985).
97. YOKOTA, T., S. WATANABE, Y. OGINO, I. YAMAGUCHI, and N. TAKAHASHI: Radioimmunoassay for Brassinosteroids and its Use for Comparative Analysis of Brassinosteroids in Stems and Seeds of *Phaseolus vulgaris*. J. Plant Growth Regul., **9**, 151 (1990).
98. TAKATSUTO, S., B. YING, M. MORISAKI, and N. IKEKAWA: Microanalysis of Brassinolide and its Analogues by Gas Chromatography and Gas Chromatography-Mass Spectrometry. J. Chromatogr., **239**, 233 (1982)
99. IKEKAWA, N., and S. TAKATSUTO: Microanalysis of Brassinosteroids in Plants by Gas Chromatography/Mass Spectrometry. Mass Spectroscopy (Japan), **32**, 55 (1984).
100. SCHMIDT, J., B. SPENGLER, B. VOIGT, and G. ADAM: New Brassinosteroids from European Cultivated Plants. Proc. 23rd Annual Meeting Plant Growth Regulator Society of America, Calgary, p. 61. 1996.
101. GOAD, L.J.: Phytosterols. In: Methods in Plant Biochemistry (P.M. DEY, and J.B. HARBORNE, eds.), p. 369. London: Academic Press. 1991.
102. GAMOH, K., T. KITSUWA, S. TAKATSUTO, Y. FUJIMOTO, and N. IKEKAWA: Determination of Trace Brassinosteroids by High Performance Liquid Chromatography. Anal. Sci., **4**, 533 (1988).
103. GAMOH, K., and S. TAKATSUTO: A Boronic Acid Derivative as a Highly Sensitive Fluorescence Derivatization Reagent for Brassinosteroids in Liquid Chromatography. Anal. Chim. Acta, **222**, 201 (1989).
104. GAMOH, K., S. TAKATSUTO, and N. IKEKAWA: Effective Separation of C-24 Epimeric Brassinosteroids by Liquid Chromatography. Anal. Chim. Acta, **256**, 319 (1992).
105. GAMOH, K., H. SAWAMOTO, S. TAKATSUTO, Y. WATABE, and H. ARIMOTO: Ferroceneboronic Acid as a Derivatization Reagent for the Determination of Brassinosteroids by High- Performance Liquid Chromatography with Electrochemical Detection. J. Chromatogr., **515**, 227 (1990).
106. GAMOH, K., M.C. PRESCOTT, L.J. GOAD, and S. TAKATSUTO: Analysis of Brassinosteroids by Liquid Chromatography-Mass Spectrometry. Bunseki Kagaku, **45**, 523 (1996).
107. PORZEL, A., V. MARQUARDT, G. ADAM, G. MASSIOT, and D. ZEIGAN: ^{1}H and ^{13}C NMR Analysis of Brassinosteroids. Magn. Res. Chem., **30**, 651 (1992).
108. ANDO, T., M. ABURATANI, N. KOSEKI, S. ASAKAWA, T. MOURI, and H. ABE: ^{13}C NMR Assignments of Brassinosteroids by Two-Dimensional Techniques. Magn. Res. Chem., **31**, 94 (1993).
109. FUNG, S., and J.B. SIDDALL: Stereoselective Synthesis of Brassinolide: A Plant Promoting Steroidal Lactone. J. Amer. Chem. Soc., **102**, 6581 (1980).
110. ISHIGURO, M., S. TAKATSUTO, M. MORISAKI, and N. IKEKAWA: Synthesis of Brassinolide, a Steroidal Lactone with Plant Growth Promoting Activity. Chem. Comm., **20**, 962 (1980).

111. THOMPSON, M.J., N.B. MANDAVA, W.J. MEUDT, W.R. LUSBY, and D.W. SPAULDING: Synthesis and Biological Activity of Brassinolide and its 22β,23β-Isomer: Novel Plant Growth Steroids. Steroids, **38**, 567 (1981).

112. ZHOU, W.-S.: Synthesis of Brassinosteroid. Pure Appl. Chem., **61**, 431 (1989).

113. LAKHVICH, F.A., V.A. KHRIPACH, and V.N. ZHABINSKII: Synthesis of Brassinosteroids – A New Class of Plant Hormones. Usp. Khim. **60**, 1299 (1991).

114. BACK, T.G.: Stereoselective Synthesis of Brassinosteroids. In: Studies in Natural Products Chemistry (ATTA-UR-RAHMAN, ed.), p. 321. Amsterdam: Elsevier Science. 1995.

115. JIANG, B., L. HUANG, W. TIAN, and W. ZHOU: Methods for Construction of Side Chain of Brassinosteroids and Application to Synthesis of Brassinosteroids. In: Studies in Natural Products Chemistry (ATTA-UR-RAHMAN, ed.), p. 245. Amsterdam: Elsevier Science. 1997.

116. KOVGANKO, N.V., and S.K. ANANICH: Advances in the Chemical Synthesis of Brassinosteroids. Chem. Nat. Comp., **33**, 389 (1997).

117. MCMORRIS, T.C.: Chemical Synthesis of Brassinosteroids. In: ref. 14, p. 69.

118. WATANABE, T., S. TAKATSUTO, S. FUJIOKA, and A. SAKURAI: Improved Synthesis of Castasterone and Brassinolide. J. Chem. Research (S), 360 (1997).

119. OKADA, K., and K. MORI: Synthesis of Brassinolide Analogs and Their Plant Growth-Promoting Activity. Agric. Biol. Chem., **47**, 89 (1983).

120. AMBERG, W., Y.L. BENNANI, R.K. CHADHA, G.A. CRISPINO, W.D. DAVIS, J. HARTUNG, K.-S. JEONG, Y. OGINO, T. SHIBATA, and K.B. SHARPLESS: Synthesis and Crystal Structure of the Cinchona Alkaloid Derivatives used as Ligands in the Osmium-Catalyzed Asymmetric Dihydroxylation of Olefins. J. Org. Chem., **58**, 844 (1993).

121. MARINO, J.P., A. DEDIOS, L.J. ANNA, and R.F. DELAPRADILLA: Highly Stereocontrolled Formal Synthesis of Brassinolide via Chiral Sulfoxide-Directed S_N2' Reactions. J. Org. Chem., **61**, 109 (1996).

122. MCMORRIS, T.C., and P.A. PATIL: Improved Synthesis of 24-Epibrassinolide from Ergosterol. J. Org. Chem., **58**, 2338 (1993).

123. SAKURAI, A. and S. FUJIOKA: Studies on Biosynthesis of Brassinosteroids. Biosci. Biotech. Biochem., **61**, 757 (1997).

124. SAKURAI, A.: Biosynthesis. In: ref. 14, p. 91.

125. SUZUKI, H., T. INOUE, S. FUJIOKA, T. SAITO, S. TAKATSUTO, T. YOKOTA, N. MUROFUSHI, T. YANAGISAWA, and A. SAKURAI: Conversion of 24-Methylcholesterol to 6-Oxo-24-methylcholestanol, a Putative Intermediate of the Biosynthesis of Brassinosteroids, in Cultured Cells of *Catharanthus roseus*. Phytochemistry, **40**, 1391 (1995).

126. FUJIOKA, S., T. INOUE, S. TAKATSUTO, T. YANAGISAWA, T. YOKOTA, and A. SAKURAI: Identification of a New Brassinosteroid, Cathasterone, in Cultured Cells of *Catharanthus roseus* as a Biosynthetic Precursor of Teasterone. Biosci. Biotech. Biochem., **59**, 1543 (1995).

127. SUZUKI, H., S. FUJIOKA, S. TAKATSUTO, T. YOKOTA, N. MUROFUSHI, and A. SAKURAI: Biosynthesis of Brassinolide from Teasterone via Typhasterol and Castasterone in Cultured Cells of *Catharanthus roseus*. J. Plant Growth Regul. **13**, 21 (1994).

128. YOKOTA, T., Y. OGINO, N. TAKAHASHI, H. SAIMOTO, S. FUJIOKA, and A. SAKURAI: Brassinolide is Biosynthesized from Castasterone in *Catharanthus roseus* Crown Gall Cells. Agric. Biol. Chem. **54**, 1107 (1990).

129. SUZUKI, H., S. FUJIOKA, S. TAKATSUTO, T. YOKOTA, N. MUROFUSHI, and A. SAKURAI: Biosynthesis of Brassinolide from Castasterone in Cultured Cells of *Catharanthus roseus*. J. Plant Growth Regul. **12**, 101 (1993).

130. FUJIOKA, S., T. INOUE, S. TAKATSUTO, T. YANAGISAWA, T. YOKOTA, and A. SAKURAI: Biological Activities of Biosynthetically-Related Congeners of Brassinolide. Biosci. Biotech. Biochem., **59**, 1973 (1995).

131. CHOI, Y.-K., S. FUJIOKA, A. HARADA, T. YOKOTA, S. TAKATSUTO, and A. SAKURAI: A Brassinolide Biosynthetic Pathway via 6-Deoxocastastrone. Phytochemistry, **43**, 593 (1996).

132. SUZUKI, H., S. FUJIOKA, S. TAKATSUTO, T. YOKOTA, N. MUROFUSHI, and A. SAKURAI: Biosynthesis of Brassinolide in Seedlings of *Catharanthus roseus*, *Nicotiana tabacum* and *Oryza sativa*. Biosci. Biotech. Biochem., **59**, 168 (1995).

133. SUZUKI, H., T. INOUE, S. FUJIOKA, S. TAKATSUTO, T. YANAGISAWA, T. YOKOTA, N. MUROFUSHI, and A. SAKURAI: Possible Involvement of 3-Dehydroteasterone in the Conversion of Teasterone to Typhasterol in Cultured Cells of *Catharanthus roseus*. Biosci. Biotech. Biochem. **58**, 1186 (1994).

134. KOLBE, A., B. SCHNEIDER, A. PORZEL, and G. ADAM: Metabolic Inversion of the 3-Hydroxy Function in Brassinosteroids. Phytochemistry, **48**, 467 (1998).

135. ADAM, G., and B. SCHNEIDER: Uptake, Transport and Metabolism. In: ref. 14, p. 113.

136. SCHNEIDER, B., and G. ADAM: A Metabolite of 24-epi-Brassinolide in Cell Suspension Cultures of *Lycopersicon esculentum*. Phytochemistry, **36**, 319 (1994).

137. YOKOTA, T., Y. OGINO, H. SUZUKI, N. TAKAHASHI, H. SAIMOTO, S. FUJIOKA, and A. SAKURAI: Metabolism and Biosynthesis of Brassinosteroids. In: ref. 12, p. 86.

138. HAI, T., B. SCHNEIDER, and G. ADAM: Metabolic Conversion of 24-epi-Brassinolide into Pentahydroxylated Brassinosteroid Glucosides in Tomato Cell Cultures. Phytochemistry, **40**, 443 (1995).

139. HAI, T., B. SCHNEIDER, A. PORZEL, and G. ADAM: Metabolism of 24-epi-Castasterone in Cell Suspension Cultures of *Lycopersicon esculentum*. Phytochemistry, **41**, 197 (1996).

140. WINTER, J., B. SCHNEIDER, D. STRACK, and G. ADAM: Role of a Cytochrome P450-Dependent Monooxygenase in the Hydroxylation of 24-epi-Brassinolide. Phytochemistry, **45**, 233 (1997).

141. KOLBE, A., B. SCHNEIDER, A. PORZEL, and G. ADAM: Metabolism of 24-epi-Castasterone and 24-epi-Brassinolide in Cell Suspension Cultures of *Ornithopus sativus*. Phytochemistry, **41**, 163 (1996).

142. KOLBE, A., B. SCHNEIDER, A. PORZEL, J. SCHMIDT, and G. ADAM: Acyl-Conjugated Metabolites of Brassinosteroids in Cell Suspension Cultures of *Ornithopus sativus*. Phytochemistry, **38**, 633 (1995).

143. KOLBE, A., B. SCHNEIDER, A. PORZEL, B. VOIGT, G. KRAUSS, and G. ADAM: Pregnane-Type Metabolites of Brassinosteroids in Cell Suspension Cultures of *Ornithopus sativus*. Phytochemistry, **36**, 671 (1994).

144. GALBRAITH, M.N., D.H.S. HORN, E.J. MIDDLETON, J.A. THOMSON, J.B. SIDDALL, and W. HAFFERL: The Catabolism of Crustecdysone in the Blowfly *Calliphora stygia*. J. Chem. Soc. Chem. Commun., 1134 (1969).

145. LACHAISE, F., and R. LAFONT: Ecdysteroid Metabolism in a Crab: *Carcinus maenas*. Steroids, **43**, 243 (1984).

146. MEUDT, W.J.: Chemical and Biological Aspects of Brassinolide. In: Ecology and Metabolism of Plant Lipids (G. FULLER, ed.), p. 53. Washington DC: American Chemical Society. 1987.

147. ADAM, G., and U. PETZOLD: Brassinosteroide – eine neue Phytohormon-Gruppe? Naturwissenschaften, **81**, 210 (1994).
148. BROSA, C.: Biological Effects of Brassinosteroids. In: Biochemistry and Function of Sterols (E.J. PARISH, and W.D. NES, eds.), p. 201. Boca Raton: CRC Press. 1997.
149. SASSE, J.: Physiological Actions of Brassinosteroids. In: ref. 14, p. 137.
150. TAKATSUTO, S., N. YAZAWA, N. IKEKAWA, T. TAKEMATSU, Y. TAKEUCHI, and M. KOGUCHI: Structure-Activity Relationship of Brassinosteroids. Phytochemistry, **22**, 2437 (1983).
151. TAKATSUTO, S., N. IKEKAWA, T. MORISHITA, and H. ABE: Structure-Activity Relationship of Brassinosteroids with Respect to the A/B-Ring Functional Groups. Chem. Pharm. Bull., **39**, 211 (1987).
152. YOKOTA, T., and K. MORI: Molecular Structure and Biological Activity of Brassinolide and Related Brassinosteroids. In: Molecular Structure and Biological Activity of Steroids (M. BOHL, and W.L. DUAX, eds.), p. 317. Boca Raton: CRC. 1992.
153. BROSA, C.: Structure-Activity Relationship. In: ref 14, p. 191.
154. BROSA, C., J.M. CAPDEVILA, and I. ZAMORA: Brassinosteroids: A New Way to Define the Structural Requirements. Tetrahedron, **52**, 2435 (1996).
155. BROSA, C., I. ZAMORA, E. TERRICABRAS, L. SOCA, R. PERACAULA, and C. RODRIGUEZ-SANTAMARTA: Synthesis and Molecular Modeling: Related Approaches to Progress in Brassinosteroid Research. Lipids, **32**, 1341 (1997).
156. STOLDT, M., A. PORZEL, G. ADAM, and W. BRANDT: Side Chain Conformation of the Growth-Promoting Phytohormones Brassinolide and 24-Epibrassinolide. Magn. Reson. Chem., **35**, 629 (1997).
157. DROSIHN, S., A. PORZEL, B. VOIGT, W. BRANDT, C. WAGNER, K. MERZWEILER, and G. ADAM: Conformational Studies of Two New Brassinosteroid Analogues with a 22,23-trans Diol Function. J. Chem. Soc., Perkin Trans. 2, (in press).
158. HAMADA, K.: Brassinolide: Some Effects for Crop Cultivations. In: Plant Growth Regulators in Agriculture, p. 188. Taipei: FFTC Book Ser. 34. 1986.
159. IKEKAWA, N., and Y.-J. ZHAO: Application of 24-Epibrassinolide in Agriculture. In: ref. 12, p. 280.
160. CUTLER, H.G.: Advances in the Use of Brassinosteroids. In: Natural and Engineering Pest Management Agents (P.A. HEDIN, J.J. MENN, and R.M. HOLLINGWORTH, eds.), p. 85. Washington DC: American Chemical Society. 1994.
161. RAMRAJ, V.M., B.N. VYAS, N.B. GODREJ, K.B. MISTRY, B.N. SWAMI, and N. SINGH: Effects of 28-Homobrassinolide on Yields of Wheat, Rice, Groundnut, Mustard, Potato and Cotton. J. Agr. Sci., **128**, 405 (1997).
162. KAMURO, Y., and S. TAKATSUTO: Practical Application of Brassinosteroids in Agricultural Fields. In: ref. 14, p. 223.
163. TAKATSUTO, S., F. FUTATSUYA, K. KOBAYASHI, and H. SATOH: Synthesis and Biological Activity of 22,23-Epoxybrassinosteroid-2,3-diacetates. Agric. Biol. Chem., **53**, 263 (1989).
164. CLOUSE, S.D.: Molecular Genetic Studies Confirm the Role of Brassinosteroids in Plant Growth and Development. Plant Journal, **10**, 1 (1996).
165. SZEKERES, M., and C. KONCZ: Biochemical and Genetic Analysis of Brassinosteroid Metabolism and Function in *Arabidopsis*. Plant Phys. Biochem., **36**, 145 (1998).
166. ALTMANN, T.: Molecular Physiology of Brassinosteroids Revealed by the Analysis of Mutants. Planta, **208**, 1 (1999).
167. CLOUSE, S.D., and K.A. FELDMAN: Molecular Genetics of Brassinosteroid Action. In: ref. 14, p.163.

168. LI, J., P. NAGPAL, V. VITARD, T.C. MCMORRIS, and J. CHORY, A Role of Brassinosteroids in Light Dependent Development of *Arabidopsis*. Science, **272**, 398 (1996).
169. CHORY, J., M. CHATTERJEE, R.K. COOK, T. ELICH, C. FANKHAUSER, J. LI, P. NAGPAL, M. NEFF, A. PEPPER, D. POOLE, J. REED, and V. VITARD: From Seed Germination to Flowering, Light Controls Plant Development via the Pigment Phytochrome. Proc. Natl. Acad. Sci. USA, **93**, 12066 (1996).
170. FUJIOKA, S., J. LI, Y.-H. CHOI, H. SETO, S. TAKATSUTO, T. NOGUCHI, T. WATANABE, H. KURIYAMA, T. YOKOTA, J. CHORY, and A. SAKURAI: The *Arabidopsis* deetiolated2 Mutant is Blocked Early in Brassinosteroid Biosynthesis. Plant Cell, **9**, 1951 (1997).
171. SZEKERES, M., K. NÉMETH, Z. KONCZ-KÁLMAN, J. MATHUR, A. KAUSCHMANN, T. ALTMANN, G.P. RÉDEI, F. NAGY, J. SCHELL, and C. KONCZ: Brassinosteroids Rescue the Deficiency of CYP90, a Cytochrome P450, Controlling Cell Elongation and De-etiolation in *Arabidopsis*. Cell, **85**, 171 (1996).
172. KAUSCHMANN, A., A. JESSOP, C. KONCZ, M. SZEKERES, L. WILLMITZER, and T. ALTMANN: Genetic Evidence for an Essential Role of Brassinosteroids in Plant Development. Plant Journal, **9**, 701 (1996).
173. MATHUR, J., G. MOLNÁR, S. FUJIOKA, S. TAKATSUTO, A. SAKURAI, T. YOKOTA, G. ADAM, B. VOIGT, F. NAGY, C. MAAS, J. SCHELL, C. KONCZ, and M. SZEKERES: Transcription of the *Arabidopsis* CPD Gene, Encoding a Steroidogenic P450, is Negatively Controlled by Brassinosteroids. Plant Journal, **14**, 593 (1998).
174. CHOE, S., B.P. DILKES, S. FUJIOKA, S. TAKATSUTO, A. SAKURAI, and K.A. FELDMAN: The *dwf4* Gene of *Arabidopsis* Encodes a Cytochrome P450 that Mediates Multiple 22α-Hydroxylation Steps in Brassinosteroid Biosynthesis. Plant Cell, **10**, 231 (1998).
175. TAKAHASHI, T., A. GASCH, N. NISHIZAWA, and N.-H. CHUA: The DIMINUTO Gene of *Arabidopsis* is Involved in Regulating Cell Elongation. Genes Devel., **9**, 97 (1995).
176. NOMURA, T., M. NAKAYAMA, J.B. REID, Y. TAKEUCHI, and T. YOKOTA: Blockage of Brassinosteroid Biosynthesis and Sensitivity Causes Dwarfism in Garden Pea. Plant Physiol. **113**, 31 (1997).
177. CLOUSE, S.D., A.F. HALL, M. LANGFORD, T.C. MCMORRIS, and M.E. BAKER: Physiological and Molecular Effects of Brassinosteroids on *Arabidopsis thaliana*. J. Plant Growth Regul., **12**, 61 (1993).
178. CLOUSE, S.D., M. LANGFORD, and T.C. MCMORRIS: A Brassinosteroid-Insensitive Mutant in *Arabidopsis thaliana* Exhibits Multiple Defects in Growth and Development. Plant Physiol., **111**, 671 (1996).
179. AZPIROZ, R., Y. WU, J.C. LOCASCIO, and K.A. FELDMAN: An *Arabidopsis* Brassinosteroid-Dependent Mutant is Blocked in Cell Elongation. Plant Cell, **10**, 219 (1998).
180. XU, W., M.M. PURUGGANAN, D.H. POLISENSKY, D.M. ANTONSIEWICZ, S.C. FRY, and J. BRAAM: *Arabidsopsis TCH4*, Regulated by Hormones and the Environment, Encodes a Xyloglucan Endotransglycosylase. Plant Cell, **7**, 1555 (1995).

(Received February 1, 1999)

Chemistry of the Neem Tree (*Azadirachta indica* A. Juss.)

A. Akhila and K. Rani
Phytochemical Technology Division, Central Institute of Medicinal and Aromatic Plants, Lucknow, India

Contents

1. Introduction . . . 48

2. Chemistry of Limonoids . . . 50
   - 2.1. Protolimonoids . . . 50
   - 2.2. Apo-Protolimonoids . . . 56
   - 2.3. Apo-Protolimonoids Derived from Loss of 4C-Atoms from the Side Chain which Possess a Hemiacetal Group . . . 59
   - 2.4. Limonoids with Intact Four Rings and a γ-Hydroxybutenolide Side Chain . . . 59
   - 2.5. Azadirone and its Natural Analogues . . . 62
   - 2.6. Homoazadirone Group . . . 69
   - 2.7. Gedunin Group . . . 70
   - 2.8. Vilasinin Group . . . 73

   C-Seco Meliacins
   - 2.9. Nimbin Group . . . 81
   - 2.10. Nimbolide Group . . . 85
   - 2.11. Nimbinene Group . . . 87
   - 2.12. Nimbolinin Group . . . 90
   - 2.13. Salannin Group . . . 93
   - 2.14. Azadirachtol Group . . . 97
   - 2.15. Meliacarpin Group . . . 100
   - 2.16. Meliacarpinin and Azadirachtinin Group . . . 100
   - 2.17. Azadirachtin Group . . . 103
   - 2.18. Azadirachtin . . . 109
     - 2.18.1. Biological Activity . . . 109
     - 2.18.2. Structure-Activity Relationships . . . 110
     - 2.18.3. Structure Determination . . . 111

2.18.4. Chemical Reactions 112
2.18.4.1. Reaction of –OH Group 116
2.18.4.1.1. Acetylation 116
2.18.4.1.2. Silylation 116
2.18.4.1.3. Methylation 116
2.18.4.2. Hydrogenation 116
2.18.4.3. Reactions of the Enol Ether Functions 118
2.18.4.4. Saponification Reactions 118
2.18.4.5. Functional Group Chemistry of Azadirachtol 120
2.18.4.6. Oxidation Reactions 120
2.18.4.7. Functional Group Chemistry of 7-Keto Azadirachtins 121
2.18.4.8. Retro-Aldol Reaction 121
2.18.4.9. Skeletal Rearrangements 124
2.18.5. Synthesis 124
2.18.5.1. Synthesis of Dihydrofuranacetal Fragment 'A' 125
2.18.5.1.1. Preparation of Prototype Coupling Fragment 125
2.18.5.2. Decalin 'B' Synthesis 125
2.18.5.3. Coupling of 'A' and 'B' Fragments 126

3. Other Compounds 127
3.1 Diterpenoids 127
3.2 Steroids and Other Triterpenoids 127
3.3 Phenolic Compounds 127
3.3.1 Flavonoids 127
3.3.2 Flavonoglycosides 129
3.3.3 Coumarins 129
3.3.4 Dihydrochalcone 130
3.3.5 Tannins 130
3.4 Carbohydrates and Proteins 130
3.5 Sulphur Compounds 130
3.6 Hydrocarbons, Acids and Esters 130

References 132

1. Introduction

Neem (Indian liliac, *Azadirachta indica* A. JUSS., synonyms *Melia azadirachta* L., family Meliaceae) a native of the Indian sub-continent, has received world-wide attention for various reasons (*189*). Teams of scientists from fields as diversified as agriculture, medicine, veterinary science, pest control, population control etc. have concentrated their attention on the therapeutic, preventive and bio-active constituents of neem (*47, 52, 54, 63, 107, 109, 115, 125, 126, 167, 296, 299, 312*). These compounds, either in pure form or in the form of extracts obtained from different plant parts, display a vast array of biological activities (*120*, *124*) such as antimalarial (*48*, *69*, *106*, *108*, *112*, *191*, *192*, *298*), antitubercular, antiviral (*216*), antiallergic, antieczemic, antiscabic, antidermatic, antidiabetic (*67*, *250*), antigingivitic, antiinflammatory (*31*, *195*), antiperiodontitic, antipyretic (*195*), antimiotitic (*235*, *307*), cardioctonic (*302*), antipyrrhoeic, antiseborrhoeic, antifeedant (*290*, *313*), antifungal (*32*, *213*, *292*), antifuruncular, bactericidal (*215*), insecticidal (*104*, *152*), larvicidal (*118*, *119*), nematicidal (*62*), piscicidal, amoebicidal, diuretic, spermicidal (*60*, *224*, *303*), vaginal contraceptive (*282*, *283*), hypoglycaemic (*180*, *202*), immunomodulatory (*310*), antimicrobial (*281*) anti-complement (*309*) activities and several others (*207*). Several toxicity studies have also been conducted on the extracts (*83*, *284*, *285*, *286*).

Chemical investigations on neem were undertaken as early as the 1890's; however, real chemical research began only in 1942 with the isolation of nimbin (*251*), nimbidin and nimbinin from neem oil (*269*). Developments after 1960 progressed more rapidly mainly because of the availability of more refined techniques of separation and structure elucidation. Preparative HPLC, Nuclear magnetic resonance (*29*, *30*, *76*, *111*, *160*), its 2D-version and X-ray analysis played pivotal roles in isolation and identification of very minor constituents from different parts of the tree (*92*, *289*). The unravelling of their high complex structural features and biogenetic interrelationships are classic examples of natural product chemistry. The chemistry of the heartwood (*190*), the trunk bark (*247*) and the pharmacology of the leaves (*167*) has been studied in detail.

The neem constituents belonging to various chemical compound classes can be broadly divided into two major sections, *i.e.* isoprenoids and non-isoprenoids. The isoprenoids are mainly diterpenoids and triterpenoids whereas glycerides, polysaccharides, sulphur compounds, flavonoids or their glycosides, amino acids and aliphatic compounds constitute the other class. The tri- and diterpenoids isolated from different tree parts and their structural formulas are listed in Tables 1 to 19. The bitterness of neem is due to the presence of limonoids which are tetranortriterpenoids (*24*, *26*, *49*, *52*, *54*); the term limonoids is derived from the compound limonin which was first obtained from the bitter principles of citrus fruits in 1841, but whose structure was only established in 1960. Limonoids occur in several plant families but the limonoids occurring in Meliaceae are also known as meliacins. Over 350 limonoids are known to-date and about 150 are known to occur in neem (*88*, *188*, *214*, *221*, *232*, *274*, *287*, *294*). In order to obtain a better understanding of the structural relationships between these they are subdivided into several groups depending upon the common features common to the groups, the name of the major compound possessing particular skeleton being assigned as the parent compound of the group. For example, all compounds having the skeleton of azadirone (**21**) are included in the azadirone group of compounds. The following is a list of the major groups of limonoids present in neem:

1. Protolimonoids
2. Apo-protolimonoids
3. Apo-protolimonoids, C_{26} (formed after the loss of side chain)
4. Limonoids with a γ-hydroxybutenolide ring
5. Azadirone group
6. Homoazadirone group
7. Gedunin group
8. Vilasinin group
9. Nimbin group
10. Nimbolide group
11. Nimbinene group
12. Nimbolinin group
13. Salanin group
14. Azadirachtol group
15. Meliacarpin group
16. Azadirachtinin and Meliacarpinin group
17. Azadirachtin group

Tables 1–17 list all limonoids from neem with their molecular formula, molecular weight, the plant part from which they have been isolated, yield (if available in the literature), m.p., specific rotation, UV, IR, ^{1}H and ^{13}C NMR, NOESY and mass spectroscopy, derivatives and their spectra, biological activity and synthesis along with all references. Scheme 1 shows a basic structure for one compound from each group and possible biogenetic interrelationships which will be discussed in more detail in the forthcoming pages.

2. Chemistry of Limonoids

2.1. Protolimonoids

The protolimonoids (Table 1), also known as protomeliacins, are considered to be biogenetic precursor of the limonoids and contain a

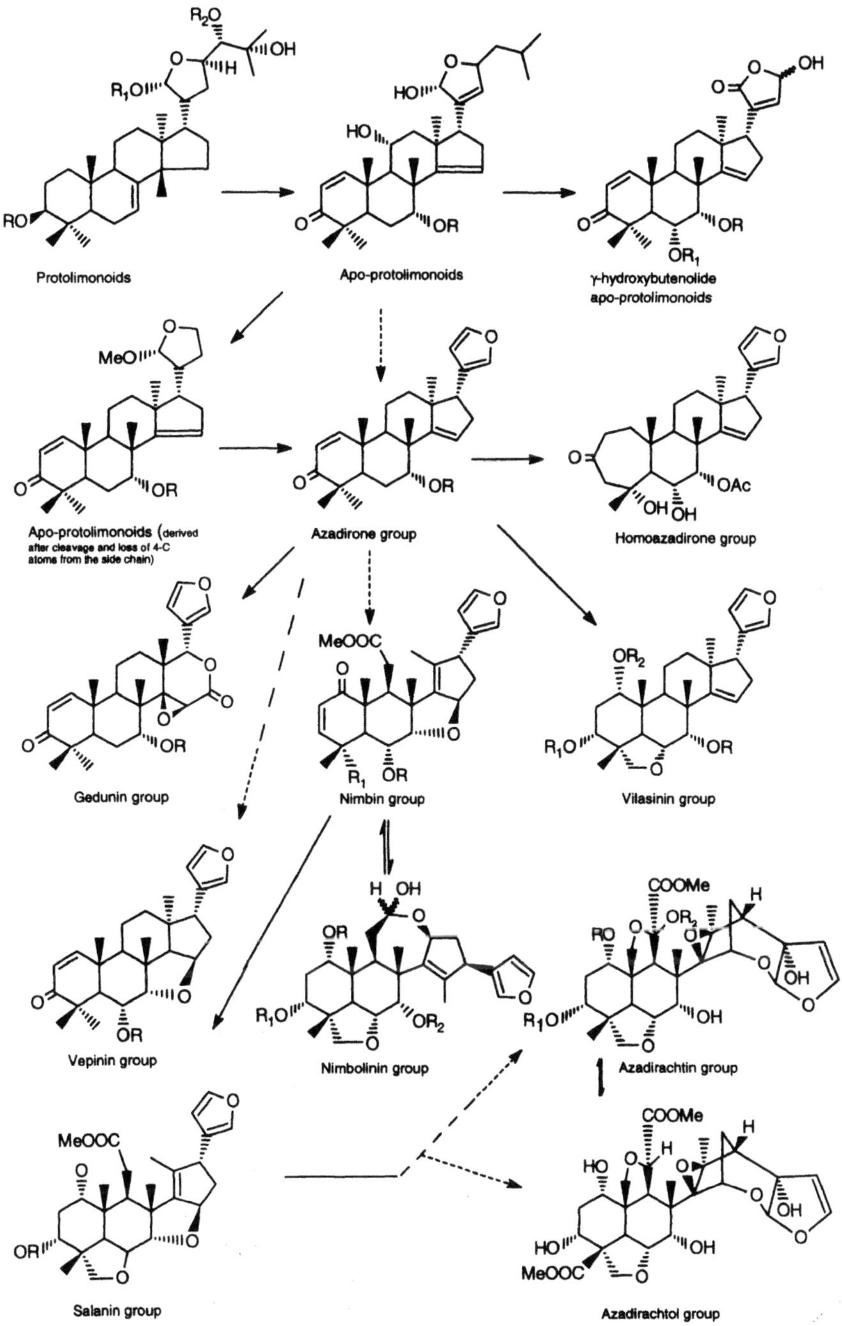

Scheme 1. Limonoids present in *A. indica* can be classified into many groups depending upon their structure. A possible biogenetic sequence leading to the different groups is illustrated

Table 1

Protolimonoids		Ref.
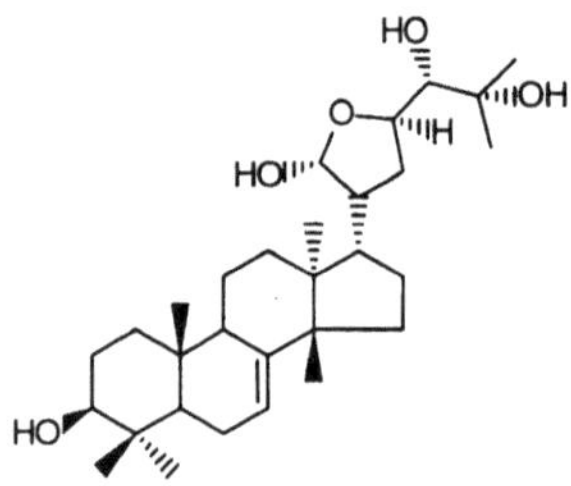 Meliantriol (**1**)	$C_{30}H_{50}O_5$; 490 **m.p.** 176–78°C; $[\alpha]_D$ −23° ($CHCl_3$, c 1.6) **Isolation**: Seed oil, leaves **Derivative**: Methyl acetate, m.p. 115–18°C, $[\alpha]_D$ −43° ($CHCl_3$, c 1.1)	*144*
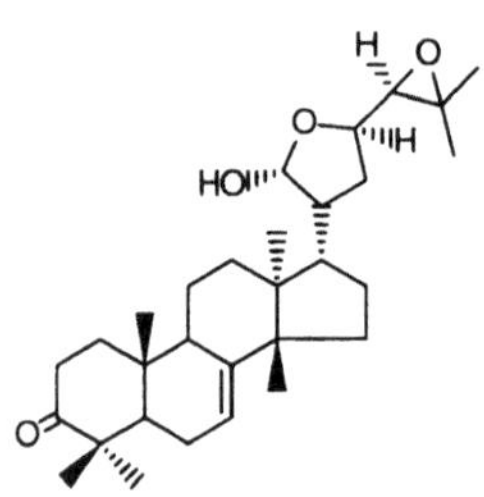 Melianone (**2**)	$C_{30}H_{46}O_4$; 470 **m.p.** 232–233°C (prismatic rods, $CHCl_3$-pentane); 223–224°C (white needles, acetone-pentane) $[\alpha]_D$ −62° (c 1.0) $\mathbf{R_f}$ 0.47 ($CHCl_3$-acetone, 9:1) **Isolation**: $CHCl_3$ extract (cold percolation) of freshly crushed fruits **Yield**: ~0.1% **Spectra**: UV, IR, ORD **Derivatives**: (i) **Acetate**, R_f 0.45 (benzene-acetate 9:1), 0.90 ($CHCl_3$-acetone 9:1), $[\alpha]_D$ −58° (c 1.0), IR; (ii) **melianone lactone**, m.p. 165–166° (acetone), $[\alpha]_D$ −48° (c 1.0), R_f 0.65° ($CHCl_3$-acetone 9:1), IR, UV; (iii) **Melianol**, m.p. 191–193°C (acetone-pentane), $[\alpha]_D$ −38° (c 1.0), R_f 0.33 ($CHCl_3$-acetone 9:1), IR; (iv) **melianol acetate**, m.p. 224–226°C (methanol), $[\alpha]_D$ −7.6° (c 1.0), R_f 0.42 (benzene-acetate 9:1), 0.77 ($CHCl_3$-acetone 9:1), IR; (v) **Epoxy-triol**, m.p. 186–188°C ($CHCl_3$-pentane), $[\alpha]_D$ −35° (MeOH, c 1.0), IR; (vi) **tetraol**. m.p. 133–135°C ($CHCl_3$-pentane), $[\alpha]_D$ −39° (MeOH, c 1.0), IR; (vii) **Melianone bromoacetal**, m.p. 176–178°C, $[\alpha]_D$ 80° (c 1.1), R_f 0.96 ($CHCl_3$-acetone 9:1), IR, ORD; **acetobromohydrin**, not crystallised, R_f 0.45 ($CHCl_3$-acetone 9:1)	*145*, *166*
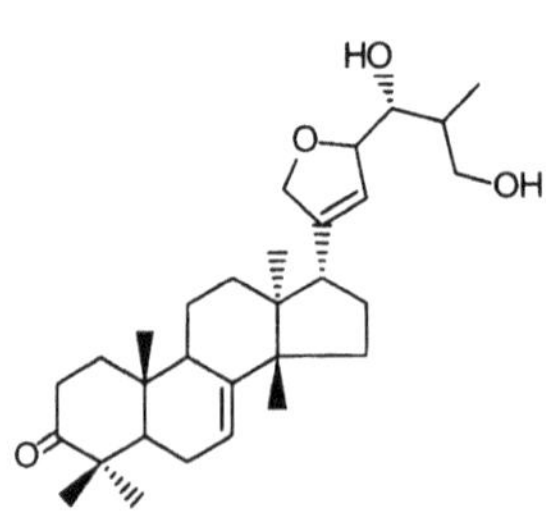 Nimbocinone (**3**)	$C_{30}H_{46}O_4$; 470.3396 **m.p.** 76–78°C (slender needle; crystallised from EtOAc) $[\alpha]_D^{23}$ +10° ($CHCl_3$) **Isolation**: Neutral fraction of EtOH extract of undried neem leaves **Yield**: 0.025% on the wt. of neutral fraction	*264*

References, pp. 132–149

Table 1 *(continued)*

Protolimonoids		Ref.
	Spectra: UV, IR, ^{1}H NMR, ^{13}C NMR **Derivatives**: (i) Diacetate, m.p. 85°C (Prismatic rods, crystallised from EtOAc), IR, ^{1}H NMR, MS. (ii) Dehydrogenation of diacetate (i), m.p.78–80°C (needles), UV, IR, MS. (iii) Reduction by $NaBH_4$: amorphous residue, UV, IR, MS, ^{1}H NMR (iv) Oxidation of reduced product: nimbocinone +24-keto nibocinone (m.p. 75–76°, UV, IR, MS) +24-keto-26-nimbocinoic acid (m.p. 88–90°, UV, IR, MS)	
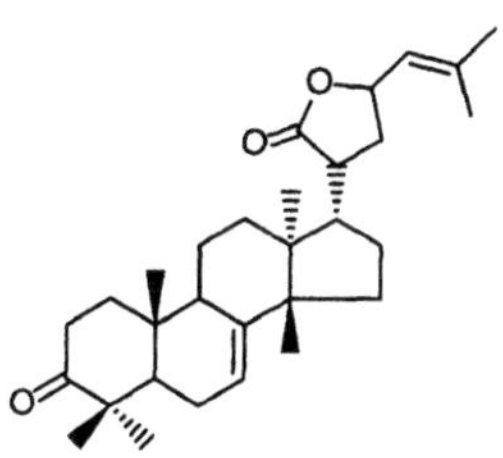 Nimolinone (**4**)	$C_{30}H_{44}O_3$; 452.3267 **m.p.** 186–90° (needles, crystallised from MeOH) $[\alpha]_D^{24}$ −24° ($CHCl_3$, c 0.25) **Isolation**: Neutral fraction of EtOH extract of fresh, uncrushed, undried ripe fruits. **Yield**: 0.14 g/20 kg fruits; 0.15% on the wt. of neutral fraction **Spectra**: UV, IR, MS, ^{1}H NMR, ^{13}C NMR, NOESY, COSY **Derivatives**: $\delta^{9,11}$-Nimolinone, m.p. 160°C (rods, crystallised from MeOH), UV, MS.	*273*
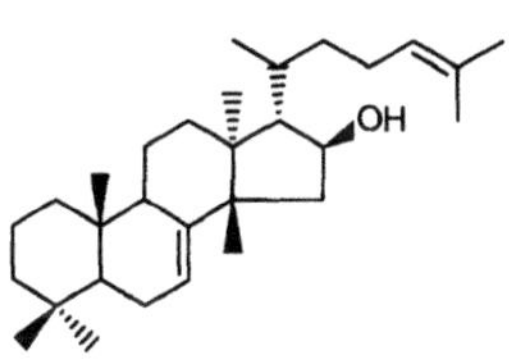 Limocinol (**5**)	$C_{30}H_{50}O$; 426.3855 Amorphous powder **Isolation**: Neutral fraction of EtOH extract of fresh, undried, uncrushed ripe fruit coatings. **Yield**: 33 mg/20 kg fruit coats; 0.017% on the wt. of neutral fraction. **Spectra**: UV, IR, ^{1}H NMR, EIMS **Derivatives**: (i) Acetate, $C_{32}H_{52}O_2$, amorphous powder, UV, IR, MS, ^{1}H NMR (ii) $\delta^{9,11}$-isomer, UV, MS	*277*
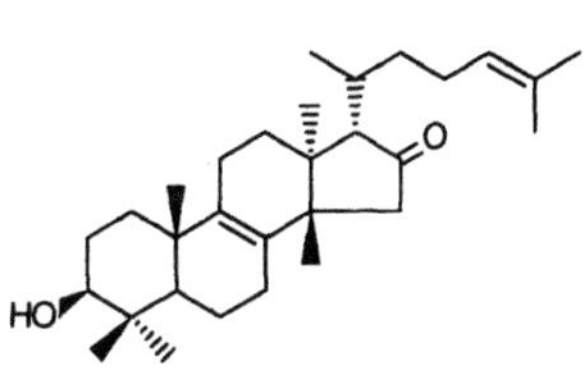 Limocinone (**6**)	$C_{30}H_{48}O$; 424.3707 amorphous powder **Isolation**: Neutral fraction of EtOH extract of fresh, undried, uncrushed ripe fruit coats. **Yield**: 40 mg/20 kg fruit coats; 0.02% on the wt. of neutral fraction. **Spectra**: UV, IR, ^{1}H NMR, ^{13}C NMR, EIMS, HRMS.	*277*

Table 1 *(continued)*

Protolimonoids		Ref.
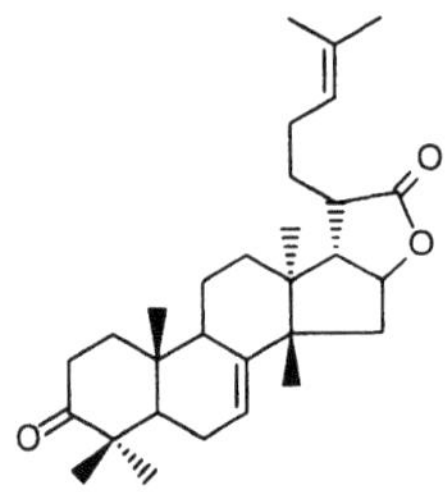 Kulactone (**7**)	$C_{30}H_{44}O_3$; 452.329 **m.p.** 163–64°C (rods) **Isolation**: Neutral fraction of EtOH extract of fresh, undried, uncrushed ripe neem fruit coats. **Yield**: 58 mg/20 kg fruit coats **Spectra**: UV, IR, ^{1}H NMR, ^{13}NMR, MS.	*278*

C8 side chain on C-17 (*149*). Meliantriol (**1**) is the first tetracyclic triterpenyl alcohol isolated from neem leaves (*144*) and is biogenetically related to 20(*S*)-tirucallol. It has been synthesised from melianone (**2**), with its structure being confirmed by synthesis. A number of related compounds such as nimbocinone (**3**) (*264*), nimolinone (**4**) (*273*), limocinol (**5**), limocinone (**6**) (*277*) kulactone (**7**) (*278*) have been isolated over the years with a biogenetic origin similar to that of euphol/tirucallol. Nimbocinone (**3**) was the first compound isolated from neem to possess hydroxy groups at C-26. The NOESY spectrum and Horeau's method successfully established the structure and absolute stereochemistry of **3** (*264*). Compounds **4**, **5**, **6** and **7** are converted to 7,9(11)-heteroannular dienes by dehydrogenation with $Hg(OAc)_2$ (*264*), thus confirming the trisubstituted nature of the 7-double bond. The stereochemistry of synthetic nimolinone (**4**) has been established by NOESY (*273*).

Biosynthesis. It is well established that squalene epoxide (**A**) on cyclisation produces an enzyme bound species (**B**) or its biogenetic equivalent (Scheme 2). This species produces lanosterol (**C**) (route 1), euphol (**D**) (Δ^8, H-20β, 20*R*, euphane skeleton) and tirucallol (**E**) (Δ^8, H-20α, 20*S*, tirucallane skeleton) (route 2) which are the biosynthetic precursors of triterpenoids and steroids. Butyrospermol (**F**), the Δ^7-isomer of euphol (route 3) formed by a double bond shift, is the expected precursor of the tetranortriterpenoids (limonoids or meliacins). Epoxidation of Δ^{24} and subsequent opening of the 24α, 25α-epoxide to a diol and introduction of oxygen at C-21, C-23 followed by formation of a cyclic hemiacetal which produces meliantriol (**1**) or its biogenetic equivalent which is expected to be a biosynthetic intermediate of other

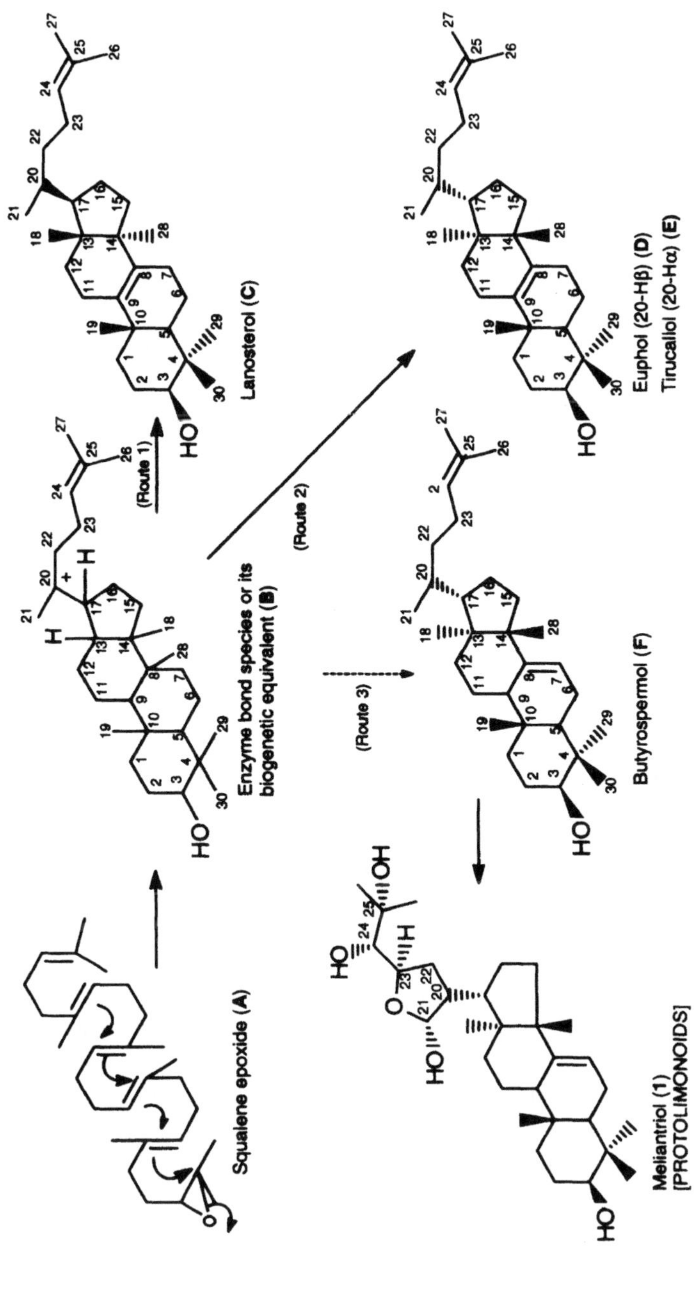

Scheme 2. Suggested biosynthetic pathway to protolimonoids from squalene epoxide

protolimonoids with all 30 carbon atoms intact such as **2**, **3**, **4**, **5**, **6** and **7**. Time course incorporation of [2-^{14}C]MVA into nimocinolide (**14**) and nimocinol (**24**) produced unequivocal evidence of the biosynthetic pathway to meliantriol (**1**) (*212*).

2.2. Apo-Protolimonoids

Apo-protolimonoids are characterized by the presence of Δ^{14} and supposed to be the immediate products formed from protolimonoids after an apo-rearrangement. Azadirachtol (**8**), also known as 20,22-dehydroazadirachnol (**8**) because of its close relation to azadirachnol (**9**) (*258*) was isolated from the fruits (*271*). One of the double bonds of **8** can be predicted on biogenetic grounds to be either at C-7 or C-14 in a butyrospremol or apo-tirucallol (apo-euphol) skeleton. However the NMR data are in good agreement with location of the double bond at C_{14}-C_{15} and with location of the acetoxy function at C-7 and α. Finally, the remaining double bond has been located at C_{20}-C_{22} because of the NMR data and formation of α,β-unsaturated-γ-lactone on oxidation (CrO_3/pyridine) (*271*). Azadirol (**10**) was isolated from neem fruit coatings (*271*). It is the first naturally occurring apotirucallol (apoeuphol) derivative with an uncyclized C-8 side chain and represents an immediate intermediate in the biosynthesis from tirucallol (euphol) of apotirucallol (apoeuphol) triterpenoids with a degraded or cyclized side chain. The ^{1}H NMR spectrum of **10** showed a pair of doublets at δ 7.13 and 5.84 ($J = 10.2$ Hz), while the ^{13}C NMR spectrum showed signals at δ158.36, 125.53 and 204.80 characteristic of ring-A 1-en-3-ones. 20, 22-Dehydroazadirachnol (**8**) and azadirachnol (**9**) are the first neem compounds with a hydroxyl group at C-11, a rather unusual position to be activated for attack by oxygen.

Biosynthesis. Biosynthesis of apo-protolimonoids is initiated from Δ^7-euphane intermediates such as butyrospermol (**F**) (Scheme 3) which undergo an apo-rearrangement induced by opening of the 7α, 8α-epoxide ring (**G**) followed by 14β-methyl migration to C-8 and formation of an α-orientated C-7 hydroxyl group thus leading to apo-protolimonoids of basic skeleton (**H**) which may or may not have an intact C-24, C-25, C-26 and C-27 side chain. These apo-protolimonoids, namely azadirachnol (**9**), 22,23-dehydroazadirachnol (**8**) and azadirol (**10**) may also be formed from meliantriol (**1**) or its biogenetic equivalent via 7α,8α-epoxide (**I**) following the steps mentioned above and shown in Scheme 3. The isolation of **8**, **9** and **10** with intact side chain indicates that the degradation of C-24, C-25, C-26, C-27 and the formation of compounds

Table 2

Apo-protolimonoids		Ref.
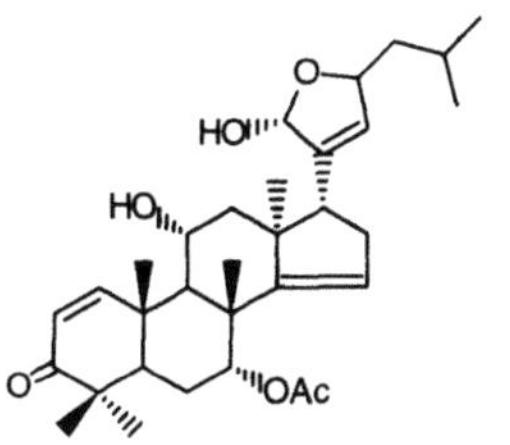 Azadirachtol or 20,22-Dehydro-azadirachnol (**8**)	$C_{32}H_{46}O_6$; 526.3289 **m.p.** 110–112°C (needles, crystallised from methanol) $[\alpha]_D^{25}$ +6.25° ($CHCl_3$) **Isolation**: EtOH extracts of fresh, unruptured ripe fruits **Yield**: 250 mg/10 kg of neem fruits **Spectra**: UV, IR, ^1H NMR, MS **Derivatives**: (i) Diacetyl azadirachtol, m.p. 120–121°C, IR, MS, ^1H NMR (ii) Tetrahydroazadirachtol (THA), m.p. 170–171°C, IR, MS, ^1H NMR (iii) $NaBH_4$ reduction of THA, m.p. 130–131°C, IR, MS (iv) Oxidation: Tri-keto derivative, m.p. 150–151°C, UV, IR, MS, ^1H NMR	*271*
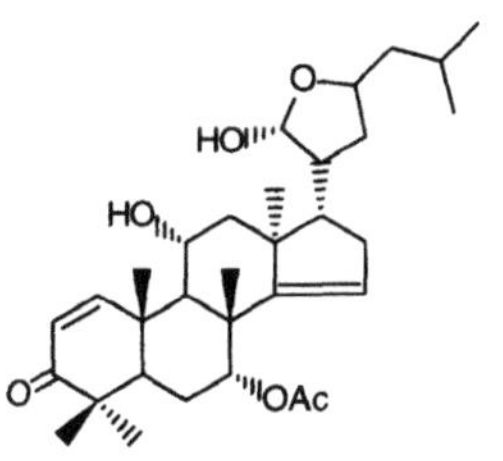 Azadirachnol (**9**)	$C_{32}H_{48}O_6$; 528.343 **m.p.** 170–171°C (needles) **Isolation**: EtOH extract of coats of fresh, unruptured, ripe fruits **Yield**: 30 mg/10 kg fruits **Spectra**: UV, IR, ^1H NMR, ^{13}C NMR, EIMS **Derivatives**: Acetate, m.p. 145–146° (needles, MeOH), UV, IR, EIMS	*258*
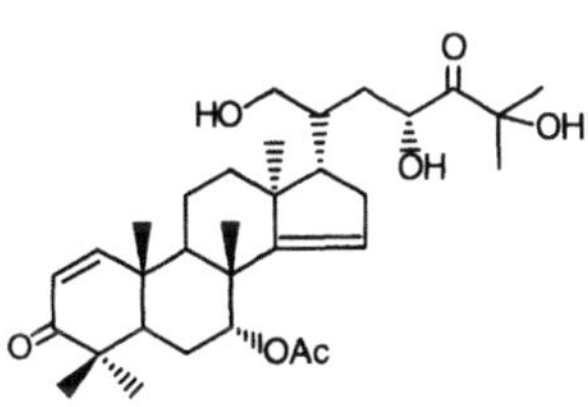 Azadirol (**10**)	$C_{32}H_{48}O_7$, 544 **m.p.** 109–112°C (white plates) **Isolation**: Neutral fraction of EtOH extract of fresh, undried, uncrushed ripe neem fruit coats. **Yield**: 0.4 g/20 kg fruit coats **Spectra**: IR, MS, ^1H NMR, ^{13}C NMR **Derivatives**: (i) Diacetyl azadirol, UV, IR, MS, ^1H NMR, ^{13}C NMR (ii) Triacetyl azadirol, UV, IR, MS, ^1H NMR, ^{13}C NMR	*278*

11, **12** and **13** with a hemiacetal ring and compunds **14**, **15**, **16**, **17**, **18**, **19** and **20** with a γ-hydroxybutenolide side chain is preceded by the apo-rearrangement. Several studies of such rearrangements, relationships among meliane-meliacins and oxidative reactions of biogenetic interest have been conducted by the groups of LAVIE (*146*, *147*, *148*) and BUCHANAN (*43*).

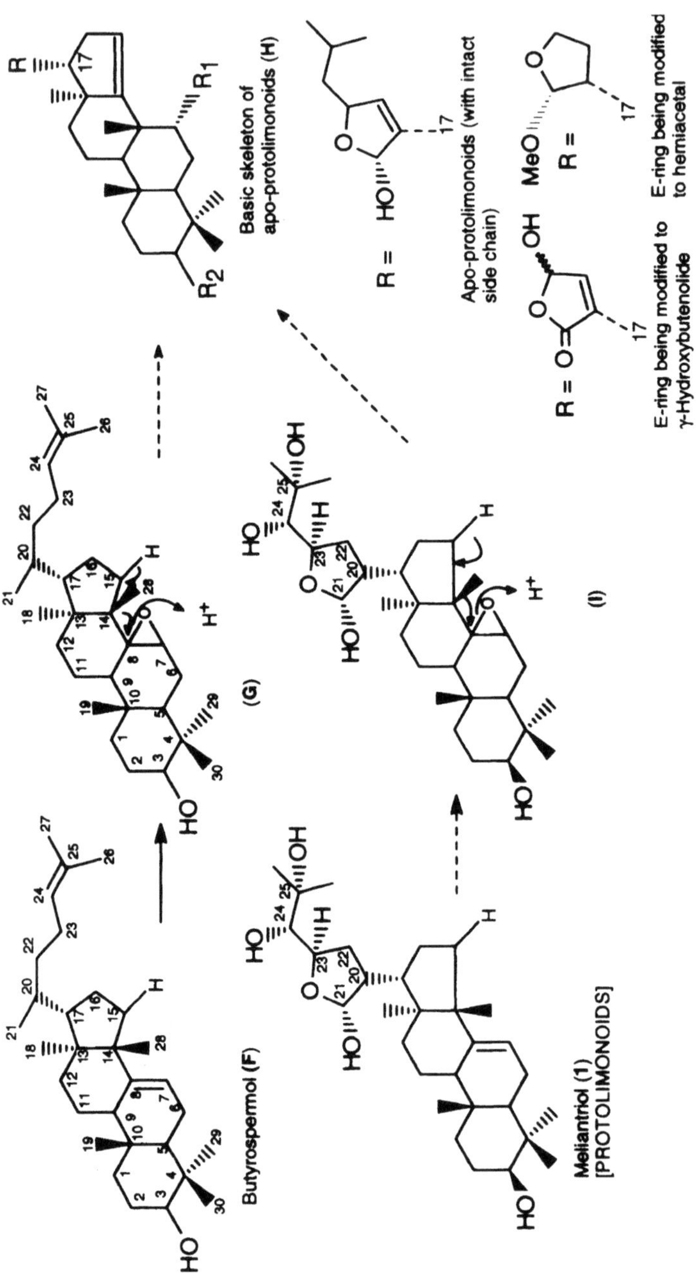

Scheme 3. Mechanism of apo-rearrangement to form apo-protolimonoids

2.3. Apo-Protolimonoids Derived from Loss of 4C-Atoms from the Side Chain which Possess a Hemiacetal Group

Limocin A (**11**), Limocin B (**12**) and limocinin (**13**) isolated from fresh, undried, ripe fruit coats (*277*) are the only tetranortriterpenoids which possess a tetrahydrofuran hemiacetal as the side chain attached at C-17.

Table 3

Apo-protolimonoids derived from loss of 4-C atoms from the side chain possessing a hemiacetal group		Ref.
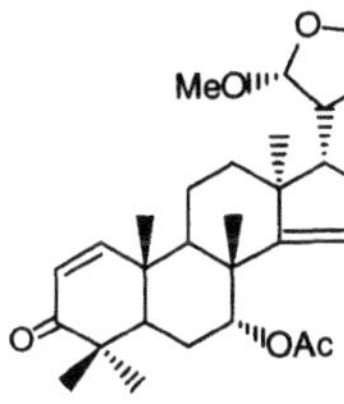 Limocin-A (**11**)	$C_{29}H_{42}O_5$; 470.3022 **Isolation**: Neutral fraction of EtOH extract of fresh, undried ripe fruit coats. **Spectra**: UV, IR, FDMS, ^{1}H NMR, ^{13}C NMR, COSY-45, NOESY	*277*
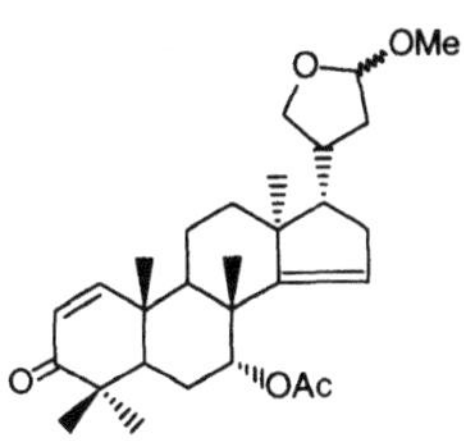 Limocin-B (**12**)	$C_{29}H_{42}O_5$; 470.3022 **Isolation**: Neutral fraction of EtOH extract of fresh, undried ripe fruit coats. **Spectra**: UV, IR, FDMS, ^{1}H NMR, ^{13}C NMR, COSY-45, NOESY	*277*
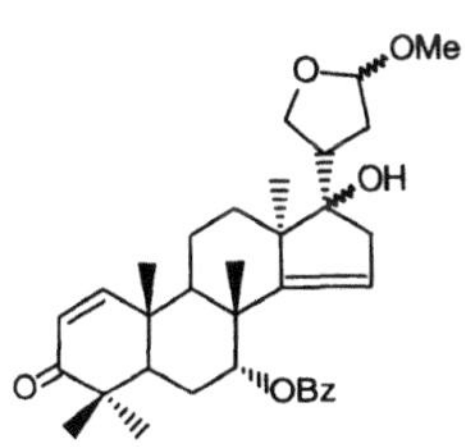 Limocinin (**13**)	$C_{33}H_{40}O_5$; 516 **Isolation**: Neutral fraction of EtOH extract of fresh, undried ripe fruit coats. **Spectra**: UV, IR, HRMS, ^{1}H NMR, ^{13}C NMR, NOESY, CD	*277*

2.4. Limonoids with Intact Four Rings and a γ-Hydroxybutenolide Side Chain

The furan ring at C-17 of proto- and apo-protolimonoids is converted into γ-hydroxybutenolides in this group of compounds. Nimocinolide

(**14**), isonimocinolide (**15**) and isonimbocinolide (**17**) have been isolated from neem leaves (*283*) and their spectral data were compared with nimocinol (**24**), which is a furan analogue of these two compounds. The noticeable chemical shifts of hydroxybutenolides in ^{1}H NMR and ^{13}C

Table 4

Limonoids with γ-hydroxybutenolide ring		Ref.
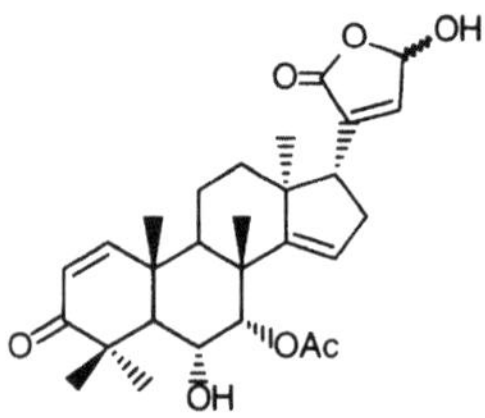 Nimocinolide (**14**)	$C_{28}H_{36}O_7$; 484 **m.p.** 160°C (needles, crystallised from $CHCl_3$) $[\alpha]_D$ 86.66° ($CHCl_3$, c 0.6) **Isolation**: Neutral fraction of EtOH extract of fresh, undried, unruptured neem leaves **Yield**: 0.68 g/40 kg leaves; 0.068% on the wt. of total neutral fraction **Spectra**: IR, UV, MS, ^{1}H NMR, ^{13}C NMR, NOESY **Derivatives**: (i) 6,23-Diacetyl nimocinolide, m.p. 72–75°C (white plates, crystallised from MeOH), $[\alpha]_D$ 25° ($CHCl_3$, c 0.04), UV, IR, MS (ii) **Oxidation**: 23-Keto compound, m.p. 98–100°C (needles, MeOH), $[\alpha]_D$ 10° ($CHCl_3$, c 0.2), UV, IR, MS	*283*
	Biosynthesis	*212*
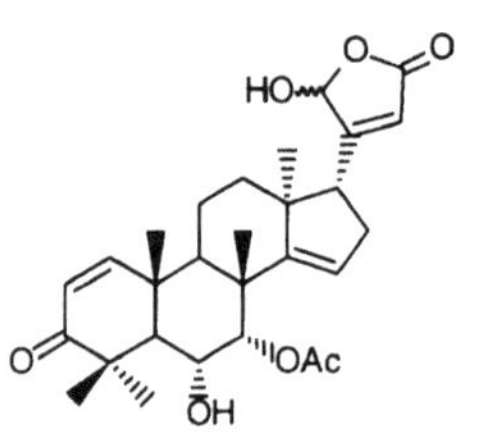 Isonimocinolide (**15**)	$C_{28}H_{36}O_7$; 484 **m.p.** 165°C (rods, crystallised from $CHCl_3$) $[\alpha]_D$ 85° ($CHCl_3$, c 0.8) **Isolation**: Neutral fraction of EtOH extract of fresh, undried, unruptured leaves **Yield**: 0.128% on the wt of neutral fraction **Spectra**: IR, UV, MS, ^{1}H NMR, ^{13}C NMR, NOESY **Derivatives**: (i) 6,23-Diacetyl nimocinolide, m.p. 88–90°C (needles, $CHCl_3$), $[\alpha]_D$ 28.57° ($CHCl_3$, c 0.07), UV, IR, MS (ii) **Oxidation**: 23-Keto compound, m.p. 98–100°C (needles, MeOH), $[\alpha]_D$ 10° ($CHCl_3$, c 0.2), UV, IR, MS	*283*
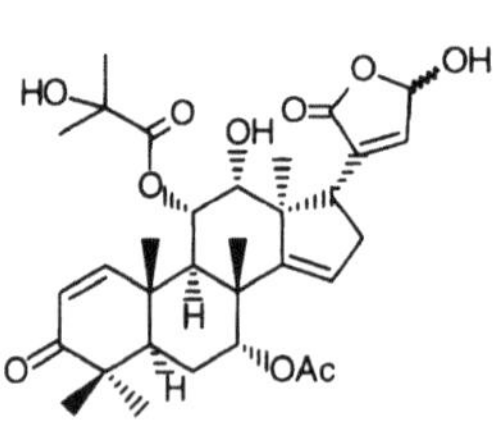 Nimbocinolide (**16**)	$C_{32}H_{42}O_{10}$; 586 **m.p.** 115–118°C (rods, crystallised from MeOH-Benzene) $[\alpha]_D^{22}$ 40.0° ($CHCl_3$) **Isolation**: Acidic fraction of EtOH extract of fresh, undried, uncrushed neem leaves **Yield**: 392 mg/40 kg leaves; 0.5% on the wt. of total acidic fraction **Derivatives**: (i) 12,23-Diacetyl nimbocinolide, m.p. 98°C (rods $CHCl_3$), IR, HRMS	*275*

Table 4 *(continued)*

Limonoids with γ-hydroxybutenolide ring		Ref.
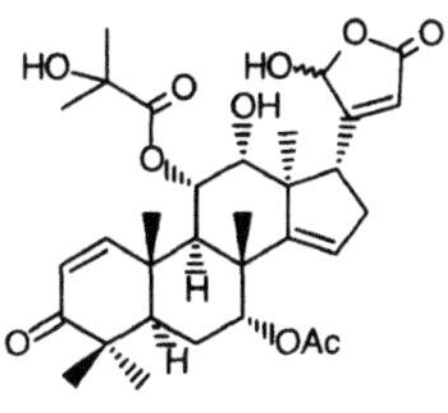 Isonimbocinolide (**17**)	$C_{32}H_{42}O_{10}$; 586.2780 **m.p.** 80–81°C (rods, $CHCl_3$) $[\alpha]_D^{22}$ 25° ($CHCl_3$) **Isolation**: Acidic fraction of ethyl acetate extract (cold percolation) of fresh, undried, uncrushed leaves **Yield**: 0.4% (on the wt. of total acid fraction) **Spectra**: IR, UV, ^{1}H NMR, ^{13}C NMR, MS **Derivative**: Acetate, m.p. 65–67°C (needles, $CHCl_3$), UV, IR, ^{1}H NMR, MS	*253*
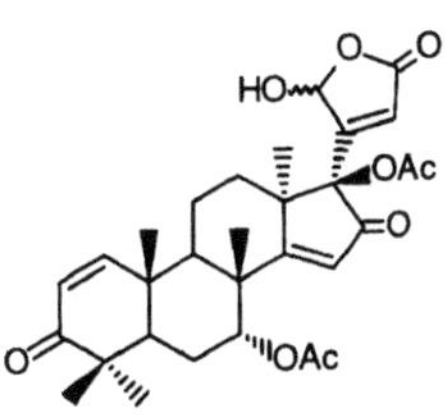 Isonimolicinolide (**18**)	$C_{30}H_{36}O_9$; 540.2399 **m.p.** 100–102°C (crystallised from EtOAc) $[\alpha]_D$ 20° ($CHCl_3$, c 0.2) **Isolation**: Neutral fraction of EtOH extract of fresh, undried, unruptured ripe fruits **Yield**: 0.02 g/20 kg fresh fruits (0.013% on dry wt basis) **Spectra**: UV, IR, MS, ^{1}H NMR, ^{13}C NMR, NOESY **Derivative**: 21-Acetyl isonimolicinolide, m.p. 110°C (plated, $CHCl_3$), $[\alpha]_D$ 15.5° ($CHCl_3$, c 0.04), IR, UV, MS	*266*
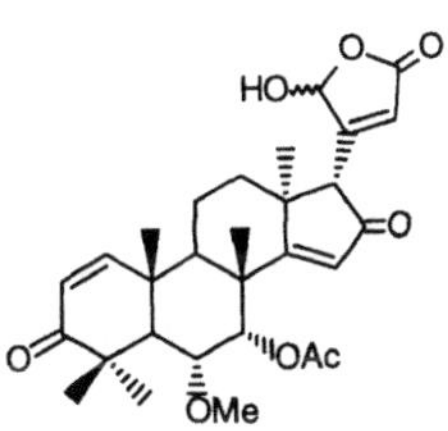 Isonimolide (**19**)	$C_{29}H_{38}O_7$; 498.2631 **m.p.** 145–148°C (irregular plates, crystallised from MeOH) $[\alpha]_D^{22}$ 50° ($CHCl_3$, c 0.04) **Isolation**: Acidic fraction of fresh, undried spring twigs **Spectra**: UV, IR, MS, ^{1}H NMR, NOESY **Derivative**: Acetyl derivative, m.p. 118°C (prismatic rods, $CHCl_3$), UV, IR, MS, ^{1}H NMR	*267, 273*
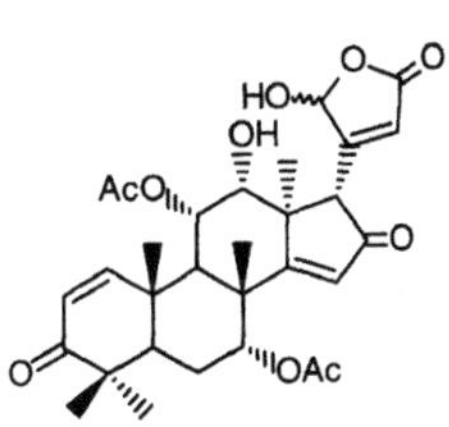 Isonimbolide (**20**)	$C_{30}H_{38}O_9$; 542.2520 **m.p.** 92–94°C (colourless plates, crystallised from $CHCl_3$) $[\alpha]_D^{22}$ 33.3° ($CHCl_3$, c 0.03) **Isolation**: Acidic fraction of fresh undried spring twigs **Spectra**: UV, IR, MS, ^{1}H NMR, NOESY **Derivative**: Acetyl derivative, m.p. 76–78°C (bunches)	*267, 273*

NMR are δ 6.9 (m) and δ 6.1 (m) for H-22 and H-23 and δ 137 (d) and δ 145 (d) for C-20 and C-22, respectively, in 21-oxo-olides. The chemical shifts in the 23-oxo-olides are δ 5.9 (m) and δ 6.0 (m) for H-21 and H-22, δ 157 (d) and δ 119 (d) for C-20 and C-22 in the ^{1}H NMR and ^{13}C NMR spectra, respectively. Nimbocinolide (**16**) and isonimbocinolide (**17**) bear oxygen functions at C-11.

2.5. Azadirone and its Natural Analogues

All four rings of the triterpenoid skeleton are intact, with the presence of functional oxygen at C-3 and C-7 being characteristic of this group of limonoids. About 8 compounds with oxygen functions at C-3 and C-7 and 15 more with an extra ketone groups at C-3 and C-16 have been isolated. Azadirone (**21**) can be isolated from neem oil (*150*) whereas nimocin (**23**) is isolated from fresh neem fruits (*254*), the only difference being the nature of the C-7 acyl function. This suggests that the precursor of both of these compounds is available in fresh fruits and finally in seeds. Nimocin (**23**) could also be a biogenetic precursor of several other benzoyl derivatives found in neem. Hydrolysis of **21** resulted in formation of 7-deacetylazadirone (**22**) which has not been isolated from neem (*124*) and it is ten times more active than **21**. Hydroxylation of **21** at C-6 might result in formation of nimocinol (**24**), also called 6-hydroxyazadirone, and its biogenetically controlled hydrogenation could lead to isomeldenin (**27**). Nimocinol (**24**) and isomeldenin (**27**) have been isolated from undried leaves (*270, 293*) and green leaves (*197*), respectively. Meldenin diol (**25**) isolated from fresh leaves (*197*) appears to be a deacetylated product of **27**. On the other hand the positions in **24** of the hydroxyl and acetyl groups at C-6 and C-7 are reversed in meldenin (**26**) which has been isolated from seed oil (*53*). A novel compound, 7-acetylneotrichilenone (**28**), has been isolated and its structure confirmed by x-ray analysis (*136*). NOE experiments had indicated that in this compound rings C and D were *cis*-fused.

Azadiradione (**29**) was reported in the methylene chloride extract of neem seeds (*150, 153*), its structure was established by ^{1}H NMR, IR, Mass and CD studies. 17-Epi-azadiradione (**30**) and 17-β-hydroxyazadiradione (**31**) have been identified in the petroleum ether extract of neem fruits (*133*); and both these compounds have been obtained in crystal line form from MeOH. The nimbidin fraction of neem contains several compounds and exhibits anti-arthritic, anti-inflammatory and anti-ulcer properties (*204, 205*). 17-Epi-nimbocinol (**33**) and nimbocinol (**34**) have been isolated from this fraction (*82*). The ^{1}H NMR spectrum of **33** showed two one proton AB doublets at δ 7.13 and 5.86 ($J = 10.2$ Hz) attributed to H_1 and H_2 of the 1-en-3-one of ring A, a singlet at δ 6.03 due to H_{15} of the 14-en-16-one of ring D, three one proton multiplets at δ 6.07 (H_{22}), 7.35 (H_{21}), 7.20 (H_{23}), a one proton singlet at δ 3.36 (H_{17}) and a three proton singlet at 1.47 (C_{13}-Me). 7-Benzoylnimbocinol (**36**), 1α-methoxy-1,2-dihydronimbinin (**39**), 1β,2β-epoxynimbinin (**40**) and 7-deacetyl-7-benzoylnimbinin (**41**) have been isolated from the neutral petroleum extract of dried seeds, their structures were confirmed by spectral studies (*136*). Nimolicinoic acid (**37**) has been isolated from the fresh, undried, unruptured ripe fruits and is the first substance with a hexanortriterpenoid apo-euphane (apo-tirucallane) skeleton and the first terpenoid acid from the neem tree (*266*). The ^{1}H and ^{13}C NMR spectra indicated that it had the same carbocyclic nucleus as that of azadiradione (**29**). Dihydronimbinin (**42**) has not been reported in neem but has been synthesised several times to prove structure of other neem compounds (*38*, *183*). Nimbinin, also known as epoxyazadiradione (**38**) (*186*) and an α,β-epoxy-δ-lactone, is presumed to originate from a precursor having an α,β-epoxy-16-ketone in the D ring by a Baeyer-Villiger type oxidation. A similar precursor with a 16-keto and 14,15-epoxy ring has been reported in this group of compounds. Its NMR spectrum showed peaks at δ 7.58 (21-H), 7.33 (23-H), 6.22 (22-H), a typical AB quartet with doublets centred at δ 7.1 and 5.8 ($J = 10$ cps) (ring A vinyl protons), a triplet centred at δ 4.66 ($J = 2.5$ cps) (7β-H), a singlet at 3.82 (17-H), sharp singlet at δ 3.3 (15-H), an acetate methyl at δ 2.01 and five tertiary methyl groups at δ 1.17, 1.15, 1.00 (2) and 0.98.

Table 5

Azadirone group		Ref.
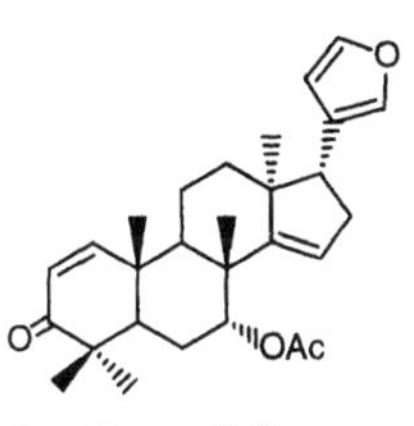 Azadirone (**21**)	$C_{28}H_{36}O_4$; 436.57 **m.p.** could not be induced to crystallisation $[\alpha]_D$ 26° (c 0.75) **Isolation**: Neem oil; R_f 0.5 (benzene-EtOAc 9:1) **Yield**: 350 mg/2 kg oil **Spectra**: IR, UV, ^{1}H NMR, CD **Derivatives**: (i) Hydrogenation: 1,2-Dihydro-azadirone, m.p. 106–110°C, $[\alpha]_D$ 6° (c 2.0), UV, IR, elemental analysis (ii) Oxidation: Azadiradione, oily, $[\alpha]_D$ −24° (c 0.9) **Synthesis**: From 7α-Acetoxy meliaca-14,20, 22-trien-3-one	*150* *42*

Table 5 *(continued)*

Azadirone group		Ref.
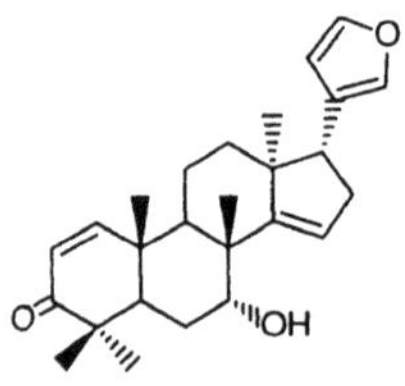 7-Deacetylazadirone (**22**)	Not found in neem. Prepared by hydrolysis from azadirone and ten times more active than azadirone	*124*
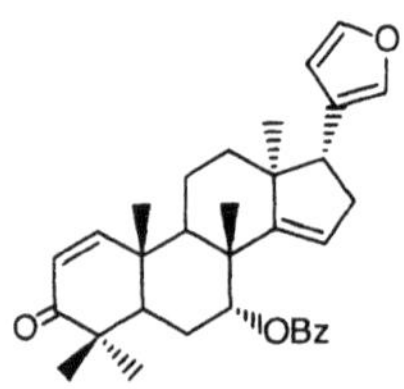 Nimocin (**23**)	$C_{38}H_{38}O_4$; 498.28 **m.p.** 190–195°C (needles, crystallised from $CHCl_3$) **Isolation**: Neutral fraction of EtOH extract of fresh, uncrushed undried ripe fruits **Yield**: 10 mg/20 kg fruits; 0.005% on the wt. of total neutral fraction **Spectra**: UV, IR, MS, ^1H NMR, ^{13}C NMR	*254*
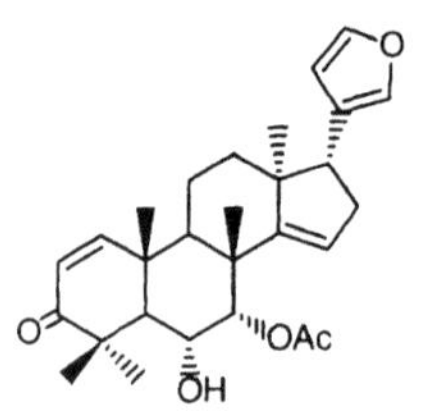 6α-Hydroxyazadirone (Nimocinol) (**24**)	$C_{28}H_{36}O_5$; 452.2585 **m.p.** 130°C (needles, crystallised from MeOH) $[\alpha]_D^{24}$ 24.4° ($CHCl_3$) **Isolation**: Neutral fraction of EtOH extract of undried leaves. **Spectra**: IR, UV, MS, ^1H NMR, ^{13}C NMR **Derivatives**: (i) 6-Acetylnimocinol, m.p. 107–108°C, $[\alpha]_D^{27}$ 75° ($CHCl_3$), IR, MS (ii) 7-Deacetylnimocinol, m.p. 100–102°C (rods, MeOH), $[\alpha]_D^{23}$ 29.12° ($CHCl_3$), IR, MS (iii) Oxidation: 6-Ketonimocinol, m.p. 227°C (needles, MeOH), $[\alpha]_D^{27}$ 27.7° ($CHCl_3$) IR, MS (iv) Isopropylidine derivative of diol (ii), m.p. 79–80°C (needles), IR, MS	*270, 293*
	Biosynthesis	*212*
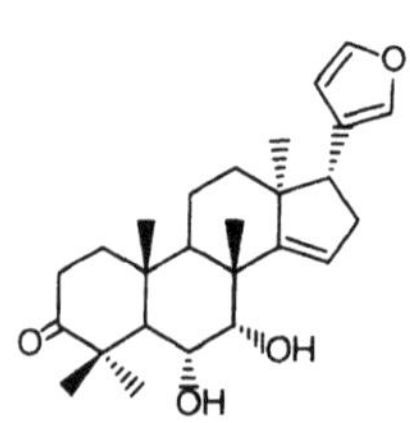 Meldenin diol (**25**)	$C_{26}H_{36}O_5$; 412.2613 **Isolation**: Fresh leaves	*196, 197*

Table 5 *(continued)*

Azadirone group		Ref.
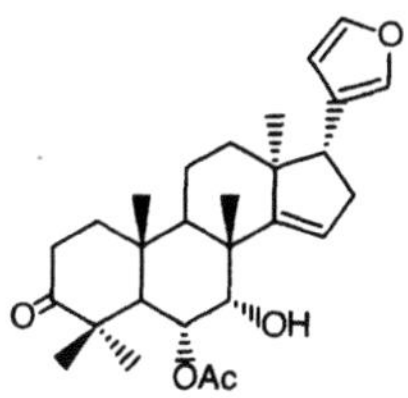 Meldenin (**26**)	$C_{28}H_{38}O_5$; 454.2719 **m.p.** 240–44°C **Isolation**: Neem seed oil **Spectra**: IR, ^{1}H NMR, ORD	*53*, *196*
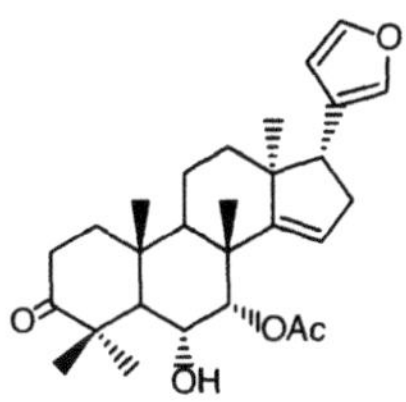 Isomeldenin (**27**)	$C_{28}H_{38}O_5$; 454.2719 **m.p.** 152°C[194a], 148–152°C[131] $[\alpha]_D$ 90° ($CHCl_3$, c 1.0) **Isolation**: n-Hexane percolation of fresh, undried, green leaves **Spectra**: ^{1}H NMR, ^{13}C NMR	*196*, *197*, *293*
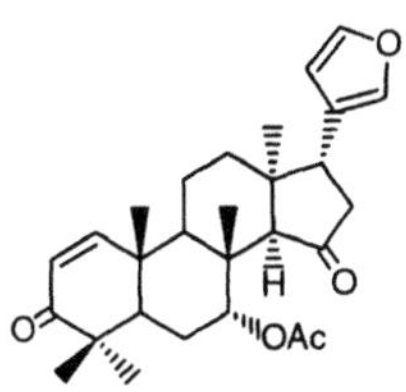 7-Acetylneotrichilenone (**28**)	$C_{28}H_{36}O_5$; 452.2548 **m.p.** 208° (crystallised from MeOH) $[\alpha]_D^{20}$ 21° ($CHCl_3$, c 1.0) **Isolation**: MeOH layer of petrol extract of dried seeds **Yield**: 700 mg/10 kg seeds **Spectra**: IR, MS, ^{1}H NMR, ^{13}C NMR	*136*
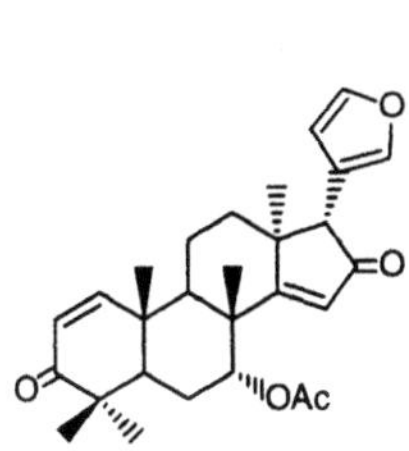 Azadiradione (**29**)	$C_{28}H_{34}O_5$; 450 **Colourless needles** (*153*), **Oily** (*150*) $[\alpha]_D$ −24° (c 0.9)	*55*, *95*, *143*, *150*, *153*
	Isolation: (i) CH_2Cl_2 extract of seeds	*153*
	(ii) By HPLC	*95*
	Yield: 620 mg/15 g CH_2Cl_2 extract **Spectra**: UV, IR, MS, ^{1}H NMR, CD	
	Derivatives: (i) Saponification: Nimbocinol, MS, ^{1}H NMR	*143*
	(ii) Hydrogenation: 1,2-Dihydroazadiradione, m.p. 178–79°C, $[\alpha]_D$ −28° (c 0.9), UV, IR	*150*
	Synthesis: Synthesized stereoselectively from *trans*, *trans*-farnesol	*55*

Table 5 *(continued)*

Azadirone group		Ref.
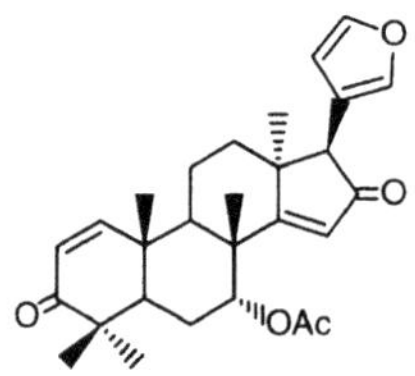 17-Epi-azadiradione (**30**)	$C_{28}H_{34}O_5$;450.6 **m.p.** 205° (crystallised from MeOH) $[\alpha]_D^{20}$ −97° ($CHCl_3$, c 1.0) **Isolation**:Petroleum ether extract of neem fruits **Yield**: 350 mg/10 kg fruits **Spectra**: IR, MS, ^{1}H NMR, ^{13}C NMR	*133*
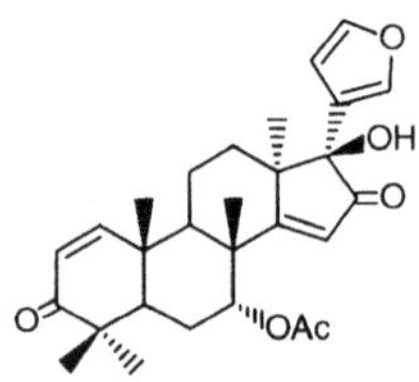 17β-Hydroxyazadiradione (**31**)	$C_{28}H_{34}O_6$; 466.6 **m.p.** 177° (crystallised from MeOH) $[\alpha]_D^{20}$ 106° ($CHCl_3$. c 1.0) **Isolation**: Petroleum ether extract of neem fruits **Yield**: 1.5 g/10 kg neem fruits **Spectra**: IR, UV, MS, ^{1}H NMR, ^{13}C NMR, CD	*133, 259*
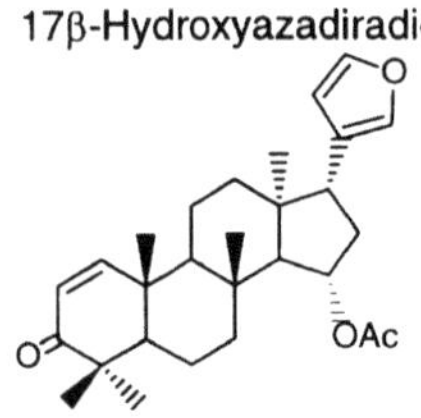 Neeflone or 15-Acetoxy-7-deacetoxydihydroazadirone (**32**)	$C_{28}H_{38}O_4$; 438 **Waxy colourless solid** **Isolation**: Light petroleum ether extract of neem flowers **Yield**: 25 mg/250 g flowers (1% of the extract) **Spectra**: UV, IR, ^{1}H NMR, ^{13}C NMR, 2D-COSY, MS	*182*
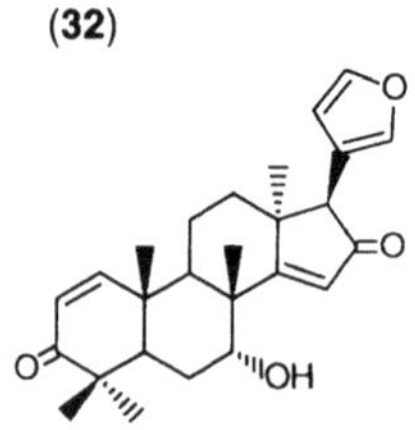 17-Epi-nimbocinol (**33**)	$C_{26}H_{32}O_4$; 408 **m.p.** 253–55°C (needles, crystallise from MeOH) $[\alpha]_D^{30}$ −72° ($CHCl_3$, c 0.5) **Isolation**: MeOH extract of Neem oil **Yield**: 0.352 g/400 g oil **Spectra**: UV, IR, MS, ^{1}H NMR, ^{13}C NMR **Derivative** Acetate	*82*
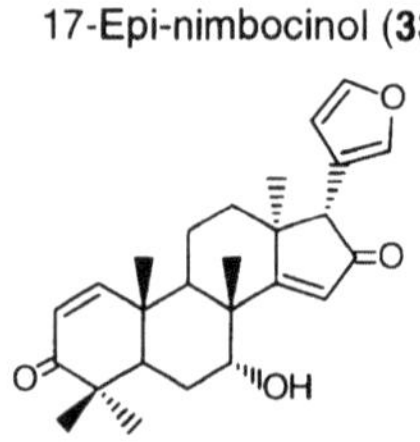 Nimbocinol (**34**)	$C_{26}H_{32}O_4$; 408 **Spectra**: ^{1}H NMR, ^{13}C NMR	*257*

Table 5 *(continued)*

Azadirone group		Ref.
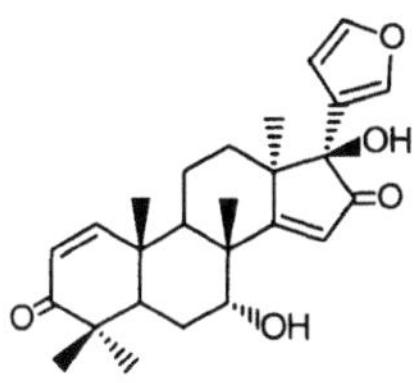 17β-Hydroxynimbocinol (**35**)	$C_{26}H_{32}O_5$; 424 Colourless needles **Isolation**: CH_2Cl_2 extract of neem seeds **Yield**: 36 mg/19 g CH_2Cl_2 extract **Spectra**: IR, UV, MS, ^1H NMR, ^{13}C NMR **Activity**: Insect growth inhibitory activity against *Heliothis virescens*	*153*
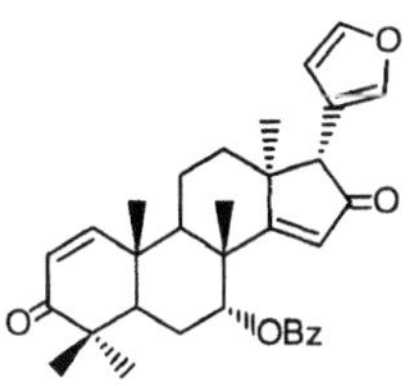 7-Benzoylnimbocinol (**36**)	$C_{33}H_{36}O_5$; 512 Amorphous power $[\alpha]_D^{20}$ 38.8° ($CHCl_3$, c 1.0) **Isolation**: Neutral extract of petrol extract of dried seeds **Yield**: 700 mg/10 kg seeds **Spectra**: IR, UV, MS, ^1H NMR, ^{13}C NMR	*136*
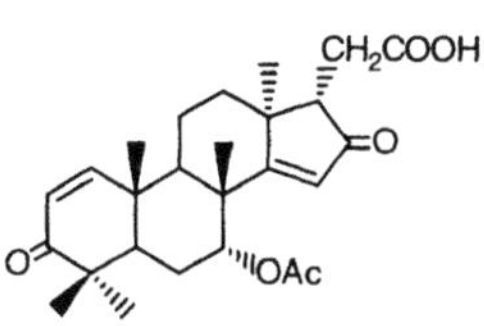 Nimolicinoic acid (**37**)	$C_{26}H_{34}O_6$; 442.2312 **m.p.** 92–94°C (needles, $CHCl_3$) $[\alpha]_D$ −14.28° ($CHCl_3$, c 0.07) **Isolation**: Neutral fraction of fresh, undried, unruptured ripe fruits **Yield**: 8 mg/20 kg fruits (0.005% on the dry wt. basis) **Spectra**: IR, UV, MS, ^1H NMR, ^{13}C NMR **Derivatives**: Methyl derivative, m.p. 104–106°C, $[\alpha]_D$ −10°, UV, IR, MS	*266*
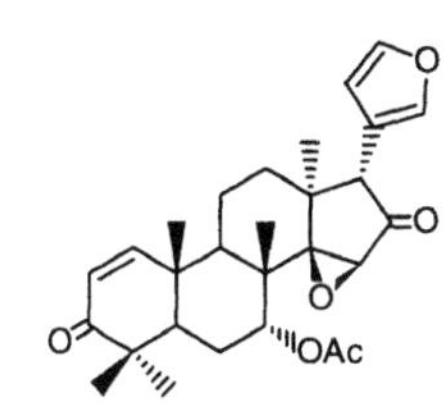 Epoxyazadiradione or Nimbinin (**38**)	$C_{28}H_{34}O_6$; 466 **m.p.** 199–203°C $[\alpha]_D$ −75° ($CHCl_3$)[99], 4.0° (c 1.0)[104], 45.0° ($CHCl_3$) **Isolation**: seed oil, R_f 0.37 (benzene-EtOAc 9:1) **Yield**: 1% of oil **Spectra**: UV, IR, MS, ^1H NMR, ^{13}C NMR, negative ion MS, CD **Reactions**	*143, 150, 168, 186* *276*

Table 5 *(continued)*

Azadirone group		Ref.
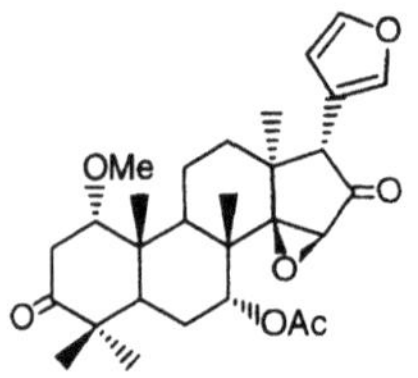 1α-Methoxy-1,2-dihydronimbinin (**39**)	$C_{29}H_{38}O_7$; 498.6 **m.p.** 235–236°C (MeOH) $[\alpha]_D^{20}$ −1.4° ($CHCl_3$, c 1.0) **Isolation**: Petrol extract of dried seeds **Yield**: 300 mg/10 kg seeds **Spectra**: IR, ^1H NMR, ^{13}C NMR, NOE, MS	*136*
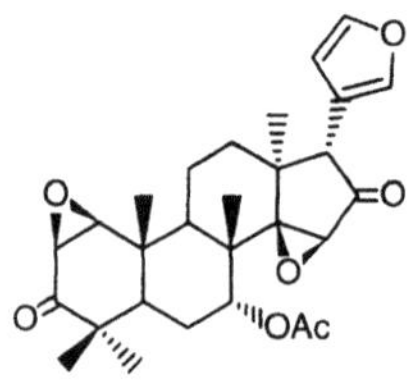 1β,2β-Epoxynimbinin (**40**)	$C_{28}H_{34}O_7$; 482.6 **m.p.** 110–111°C $[\alpha]_D^{20}$ −3.5° ($CHCl_3$, c 1.0) **Isolation**: Seed Oil **Yield**: 300 mg/10 kg seeds **Spectra**: IR, MS, ^1H NMR, ^{13}C NMR	*136*
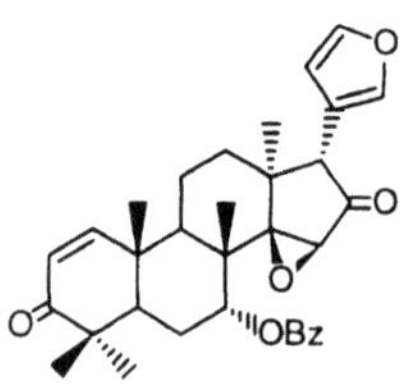 7-Deacetyl-7-benzoylnimbinin (**41**)	$C_{36}H_{36}O_6$; 528 **amorphous powder** $[\alpha]_D^{20}$ 81.4° ($CHCl_3$, c 1.0) **Isolation**: Petrol extract of dried neem seeds **Yield**: 1.2 g/10 kg seeds **Spectra**: UV, IR, ^1H NMR, ^{13}C NMR, MS	*136*
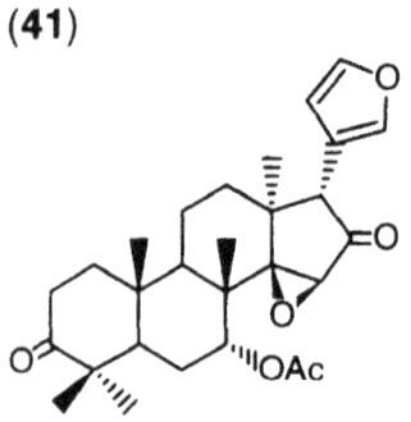 Dihydronimbinin (**42**)	Not reported in neem	*38, 183*

Biosynthesis. The azadirone group of compounds consists of limonoids which contain all the four rings of basic skeleton and a furan ring at C-17 with α-orientation. A possible intermediate (**J**) (Scheme 4) has been postulated which could have been formed by an apo-rearrangement of meliantriol (**1**) or its biogenetic equivalent. Elimination of

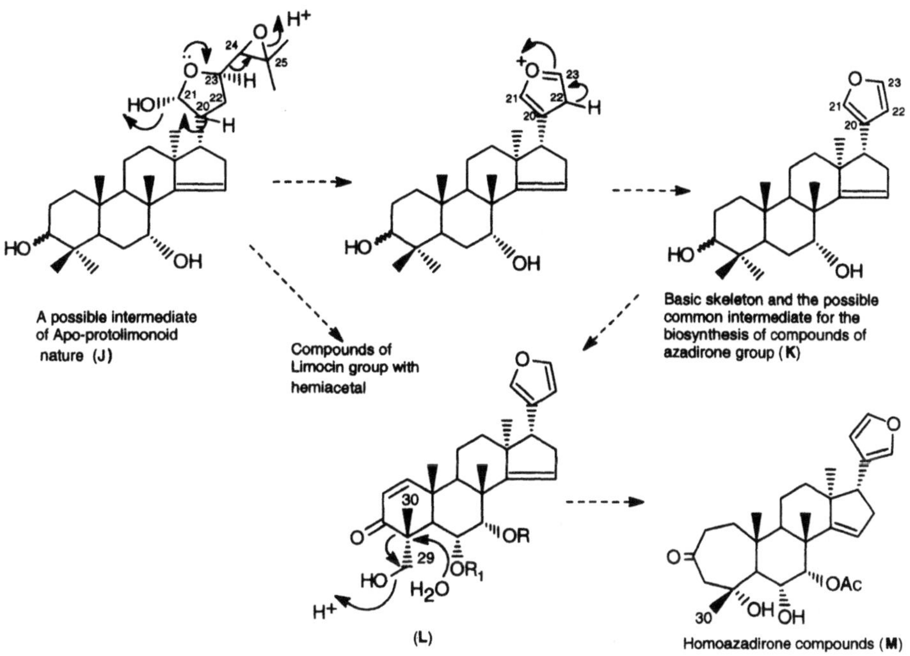

Scheme 4. Suggested mechanism for the elimination of four side chain carbon atoms to form compounds of the azadirone and homoazadiradione groups

H_2O from C-20-C-21 (dehydration) and the presence of a lone pair of electrons on oxygen which enforces the elimination of C-24, C-25, C-26, C-27 (side chain) as shown in Scheme 4 forms the basic skeleton (**K**) of azadirone compounds. Oxidation of C-29-methyl (**L**) and attack of H_2O at C-4 opens the ring and C-29 is absorbed in ring A to convert the six-membered ring A into a seven-membered one, thus producing the basic skeleton (**M**) for homoazadirone compounds **43** and **44**.

2.6. Homoazadirone Group

Compounds of this group have a seven-membered A ring, with one of the two methyls at C-4 being merged into the A ring. A novel homologue of azadirone has been isolated from dried leaves and named 4α,6α-dihydroxy-A-homoazadirone (**43**) (*41*). The structure of this compound has been assigned on the basis of 2D, ^{1}H & ^{13}C NMR and NOE studies. 1,2-Dihydro-4α,6α-dihydroxy-A-homoazadirone or 4α-hydroxy-A-homoisomeldenin (**44**) has been isolated by Kraus *et al.* in unpublished work.

Table 6

Homo-azadirone group		Ref.
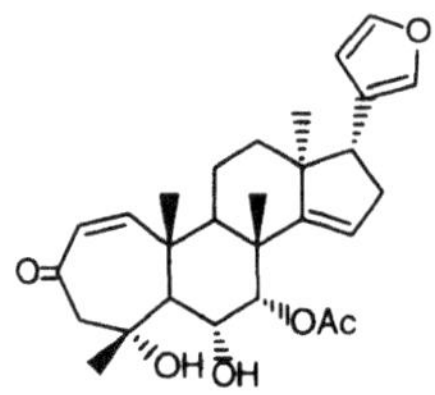 4α,6α-Dihydroxy-A-homo-azadirone (**43**)	$C_{28}H_{36}O_6$; 468 **m.p.** 177–180°C (MeOH) $[\alpha]_D^{20}$ −75° ($CHCl_3$, c 0.1) **Isolation**: Ether extract of dried leaves **Yield**: 120 mg/10.5 kg dried leaves **Spectra**: IR, ^1H NMR, ^{13}C NMR, MS	*41*
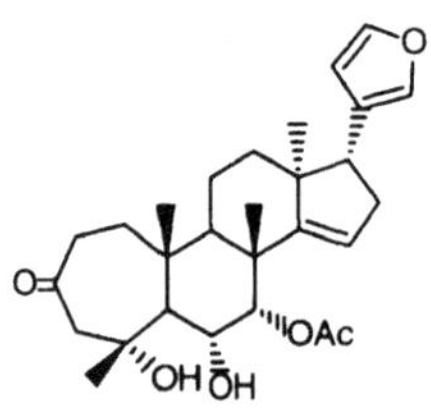 1,2-Dihydro-4α,6α-dihydroxy-A-homoazadirone or 4α-Hydroxy-A-homo-isomeldenin (**44**)	Unpublished result (W. KRAUS and A. BRUHN)	

2.7. Gedunin Group

This group consists of compounds in which the D-ring has undergone oxidative expansion through a Baeyer-Villiger type reaction. This could occur in epoxy-azadiradiones such as nimbinin (**38**). Other compounds of the azadiradione group such as nimbocinol (**34**), 17β/α-hydroxyazadiradiones (**35**) could also give rise to D-ring expansion. Gedunin (**45**) and 7-deacetylgedunin (**46**) have been found in neem bark and seed (*143, 150*) whereas 7-deacetyl-7-benzoylgedunin (**47**) has been reported only in dried seeds. Mahmoodin (**49**) belongs to the deoxygedunin group. The presence of the C-17 glycolyl side chain has been established by NMR and high resolution mass spectroscopy (HRMS). A peak at m/z 370.2127 ($C_{23}H_{30}O_4$) arising from a retro-*Diels-Alder* cleavage around ring C, and the base peak at m/z 328.2029 ($C_{21}H_{28}O_3$) resulting from the subsequent loss of a ketene molecule were significant. The absolute stereochemistry was established through NOESY spectral analysis and CD spectral data (*258*). It was postulated that, its biogenesis might occur from isonimolicinolide (**18**) through oxidation of ring D to a δ-lactone and transformation of the acetoxy group to $-OCH_2CH_2OH$. Azadirinin (**50**) has been isolated from root bark (*19*) and its structure established by ^1H NMR and a COSY-45 plot which showed through-bond connectivities of both H-1 and H-3 with H-2α, H-2β; H-5 with H-6; H-9 with both H-11α and H-11β and both of these in turn connected with H-12α and H-12β; H-22 with H-17, H-21 and H-23; and H-2′ with H-3′. These observations, and particularly the signals of H-15 and H-17, led to the conclusion that ring D is oxidized to a lactone as observed in 14,15-deoxy-gedunin derivatives (*19*). Another compound in this group is nimolicinol (**48**) isolated from fresh, ripe, undried fruits; its structure was elucidated as 17α-hydroxy-14,15-deoxy-17-epi-gedunin by spectral studies (*256*).

Table 7

Gedunin group		Ref.
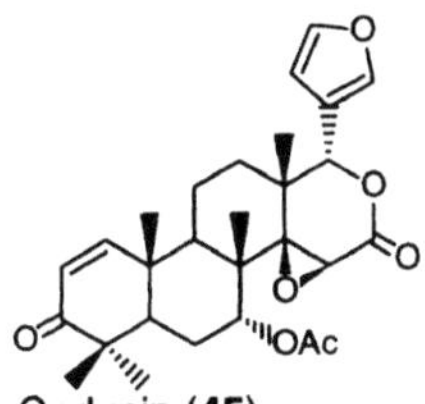 Gedunin (**45**)	$C_{28}H_{34}O_7$; 482.55 **m.p.** 218–220°C (MeOH) $[\alpha]_D^{20}$ −44° ($CHCl_3$, c 1.1) **Isolation**: (i) Seed oil (ii) 70% MeOH extract of bark R_f 0.31 (benzene-EtOAc 9:1) **Spectra**: UV, IR, ^1H NMR, ^{13}C NMR, COSY, COLOC, negative ion MS **Derivative:** (i) Hydrolysis: 7-Desacetylgedunin, 259–262°C, $[\alpha]_D$ 75° (c 0.9), UV, IR (ii) Oxidation of 7-desacetylgedunin: Oxo-derivative, m.p. 263–264°C, $[\alpha]_D$ −50° (c 1.3), UV, IR **Activity**: Antimalarial (*112*), antifungal (*213*)	*112, 143, 150, 168*
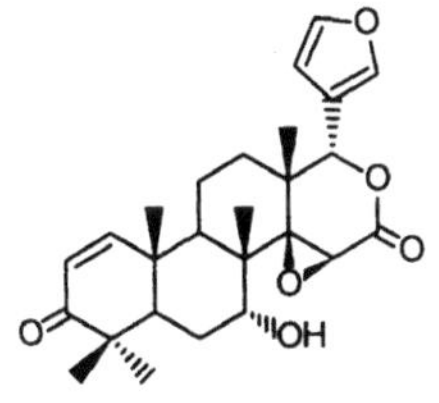 7-Deacetylgedunin (**46**)	$C_{26}H_{32}O_6$; 440.52 **m.p.** 259–262°C $[\alpha]_D$: 75° (c 1.3) **Isolation**: Seed oil R_f: 0.65 ($CHCl_3$-acetone 9:1) **Spectra**: UV, IR, elemental analysis, MS **Derivative**: Oxidation: Oxo-derivative, m.p. 263–264°C, $[\alpha]_D$ −50° (c 1.3), UV, IR	*143, 150*

Table 7 *(continued)*

Gedunin group		Ref.
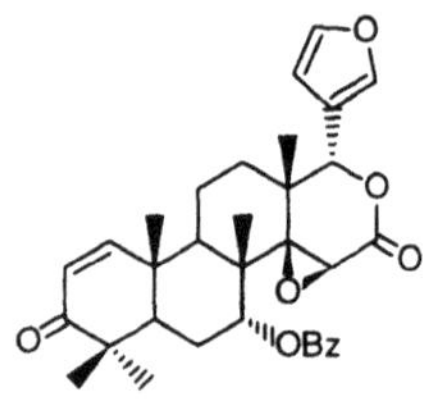 7-Deacetyl-7-benzoylgedunin (**47**)	$C_{33}H_{36}O_7$; 544 **m.p.** 278°C (MeOH) $[\alpha]_D^{20}$ 105° ($CHCl_3$, c 1.0) **Isolation**: Petroleum ether extract of seeds **Yield**: 1.5 g/10 kg of dried seeds **Spectra**: UV, IR, ^{1}H NMR, ^{13}C NMR, MS	136
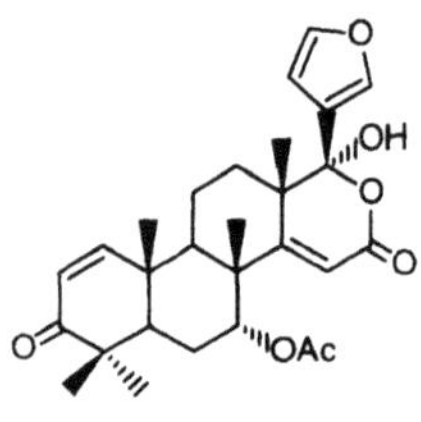 Nimolicinol (**48**)	$C_{28}H_{34}O_7$; 482 **m.p.** 270–274°C (needles, EtOAc-MeOH) **Isolation**: Neutral fraction of ethanol extract of fresh, unruptured, ripe fruits **Yield**: 1 g/20 kg fruits (0.01% on dry wt. Basis) **Spectra**: UV, IR, ^{1}H NMR, ^{13}C NMR, MS, NOE	*256*
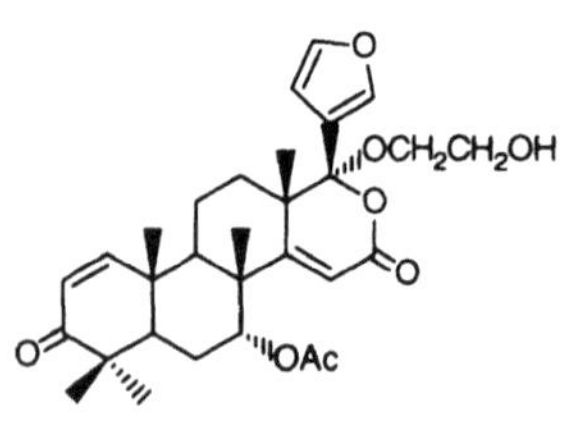 Mahmoodin (**49**)	$C_{30}H_{38}O_8$; 526.2554 **amorphous** **Isolation**: Seed oil **Yield**: 50 mg/1 lit. oil **Spectra**: IR, ^{1}H NMR, ^{13}C NMR, FDMS **Derivative**: Acetate, amorphous, UV, IR, EIMS	*258*
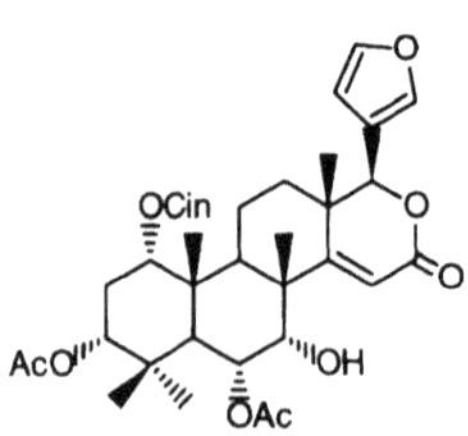 Azadirinin (**50**)	$C_{39}H_{46}O_{10}$; 674 **m.p.** amorphous powder $[\alpha]_D^{24}$ 48° ($CHCl_3$, c 6.4) **Isolation**: Neutral fraction of CH_2Cl_2 extract of root bark **Yield**: 78.6 mg/28 kg root bark **Spectra**: UV, IR, ^{1}H NMR, ^{13}C NMR, NOESY, HET-COSY, MS	*19*

Biosynthesis. Azadirone (**21**) or a compound with similar skeleton (**K**) could be metabolised to the vilasinin and gedunin group of compounds as shown in Scheme 5. Oxidation of C-29 followed by nucleophilic attack of the C-6 hydroxyl on C-29 (**N**) would produce an ether

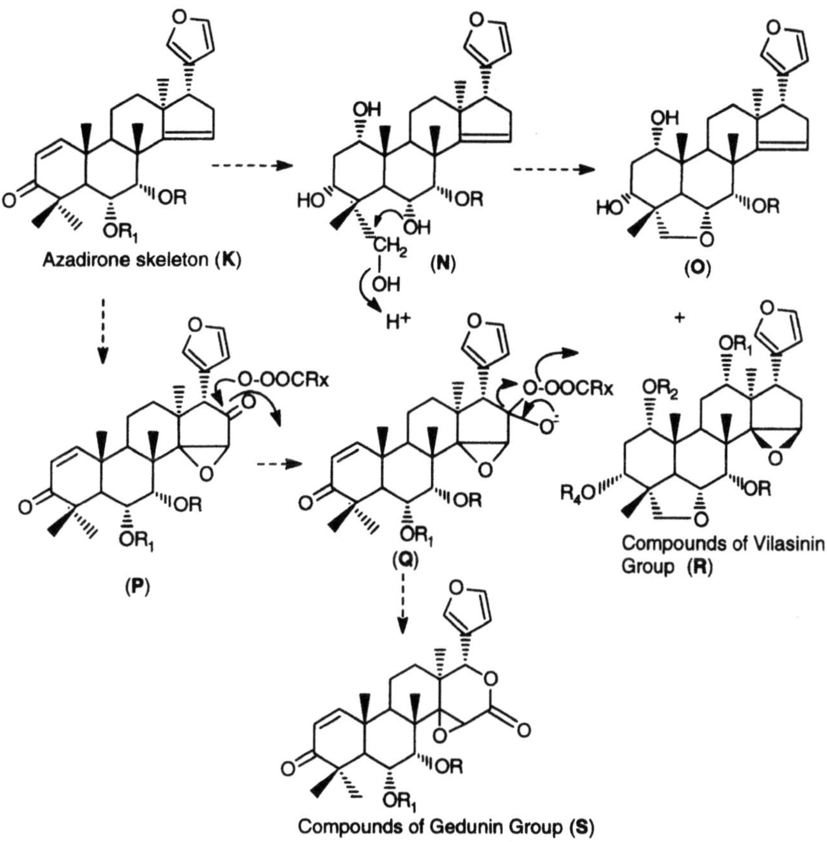

Scheme 5. A possible biosynthetic route to compounds of the vilasinin and gedunin groups

linkage between C-29 and C-6 to produce skeleton (**O**) whereas epoxidation of $\Delta^{14,15}$ would produce compounds of the vilasinin group (**R**). A Baeyer-Villiger oxidation of D-ring would insert an oxygen in ring D in between C-16 and C-17 to form compounds of the gedunin group (**S**).

2.8. Vilasinin Group

About 21 constituents are included in this group and are biosynthetic intermediates for ring C-seco limonoids. A C-7 oxygen bridged to C-15 gives rise to a B/D ring fusion in vepinin (**74**) (*187*), this type of ring fusion being characteristic of the salannin and nimbin group of compounds. **74** could be a metabolite of nimocinol (**24**). On the other hand oxidation of the C-28 methyl and formation of C-28/C-6 ether linkage gives rise to vilasinin (**51**) (*197*), 1,3-diacetylvilasinin (**52**) (*134*), and vilasinin triacetate (**53**) (*197*). Similarly esterification with either tiglic acid, cinnamic acid or senecioic acid at the C-1, C-3 and C-7 hydroxy groups produce a range of compounds such as 1-tigloyl-3-acetylvilasinin (**54**) (*121*), 1-senecioyl-3-acetylvilasinin (**55**) (*131*), 1,3,-diacetyl-7-cinnamoyl-vilasinin (**56**), also called nimbolin A (*71*), and 1-acetyl-7-tigloylvilasinin (**57**) (*121*). Oxidation of C-12 in C-ring produces 1-tigloyl-3-acetyl-12α-acetoxyvilasinin (**58**), 1,3-diacetyl-12α-acetoxyvilasinin (**59**) and 7-tigloyl-12α-acetoxyvilasinin (**60**) (*125*, *131*). The antifeedant activity of the above vilasinins has been shown to be reduced by functionalisation of C-12 (*125*). A new tetranortriterpene nimbidinin (**61**) (*174*) from the neutral fraction of nimbidin, the amorphous bitter principle of neem seed kernel, was shown to be biogenetically related to salannin. On acetylation (Py/Ac_2O) nimbidinin (**61**) yielded a triacetate showing the presence of three hydroxy groups. It yielded a monoxime showing the presence of one carbonyl group. The presence of a β-substituted furan ring attached to C-17, a common feature of meliacins, in nimbidinin (**61**) was evident from the NMR signals at δ 6.62 (1H) and 7.42 (2H) ppm. Nimbidinin appears to be a intermediate for salannin (**107**). This is supported by the co-occurrence of **61** and nimbidic acid (**89**) (*173*, *174*), the acid related to salannin, in *A. indica*. It appears that a saturated ring A with oxygen functions at C-1 and C-3 together with ether linkages between C-28/C-6 and C-7/C-15 are responsible for the antifeedant activity. 1-Acetyl-7-tigloylnimbidinin (**62**) has been reported from seeds (*125*, *131*). **61** and other vilasinin derivatives containing C-12 oxygen functions can be considered as biogenetic precusors of C-secomeliacins. Limbocinin (**63**) and Limbocidin (**64**) (*261*) are highly oxidized compounds of this group in which C-16 has been oxidized to a keto group, an epoxide is formed from $\Delta^{14,15}$, and the furan ring becomes a maleic anhydride. 1,7-Ditigloyl-3-acetylvilasinin (**73**) has been reported from *Melia composita* (*288*). Though not found in *A. indica* this compound is very similar to the compounds with the vilasinin skeleton, hence mention has been made here. Its ^{1}H NMR spectrum closely resembles 1-tigloyl-3-acetylvilasinin (**54**) except that the tiglate protons 1.30 (m, $CH_3 \times 2$), 6.92 (q, $J = 7.6$ Hz, olefinic-H × 2) and 2.05 (s, OAc × 2) were double in number. Azadirachtolide (**71**) and deoxyazadirachtolide (**72**) have been isolated from air-dried leaves (*208*) and their structures established by ^{1}H-NMR, Mass and FT-IR. The NMR spectrum of azadirachtolide (**71**) showed resonances for a senecioyloxy substituent [δ 1.88 (3H, d, $J = 1.2$ Hz), δ 2.20 (3H, d, $J = 1.2$ Hz), δ 5.70 (1H, s, br)] and acetate [δ 1.96 (3H,s)], four additional methyl singlets (δ 0.97, δ 1.0, δ 1.05, δ 1.19), two olefinic hydrogens [δ 5.50 (dd, $J = 1.84$, 3.0 Hz), 5.70 (s, br)], methylene hydrogens bonded to oxygenated carbons [δ 4.4 (1H, t, $J = 7.7$ Hz), δ 3.95 (1H, t, $J = 7.7$ Hz), δ 3.63 (1H, d, $J = 7.6$ Hz), δ 3.56 (1H, d, $J = 7.6$ Hz) and methine hydrogens bonded to oxygenated carbons. The structure of deoxyazadirachtolide (**72**) has also been established using NMR, NOESY and COSY spectroscopy (*208*). Vilasinin lactones and lactols (**65**, **66**, **67**, **68** and **69**) have also been reported in neem (*35*, *125*).

A C-19/C-29 ether linkage has been found in azadirachtanin (**75**) which has been isolated from fresh leaves (*205*); however there is no other ether or epoxide linkage between C-28/C-6, C-7/C-15 or $\Delta^{14,15}$ respectively. This is the only compound in neem with a C-19/C-29 ether bridge.

Table 8

Vilasinin group		Ref.
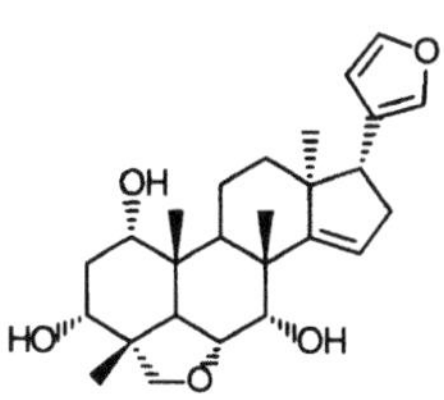 Vilasinin (**51**)	$C_{26}H_{36}O_5$; 428.2562 **m.p.** 255°C **Isolation**: green leaves	*197*
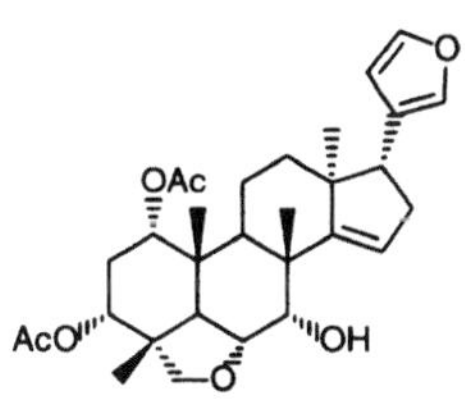 1,3,-Diacetylvilasinin (**52**)	$C_{30}H_{40}O_7$; 512.2763 **m.p.** 157–158°C (Ethyl acetate) $[\alpha]_D^{20}$ −6.5° ($CHCl_3$, c 1.0) **Isolation**: Seed oil **Spectra**: IR, ^1H NMR, ^{13}C NMR, NOE **Derivative**: 1,3,7-Triacetylvilasinin, m.p. 228°C (MeOH), IR, ^1H NMR, ^{13}C NMR	*134*
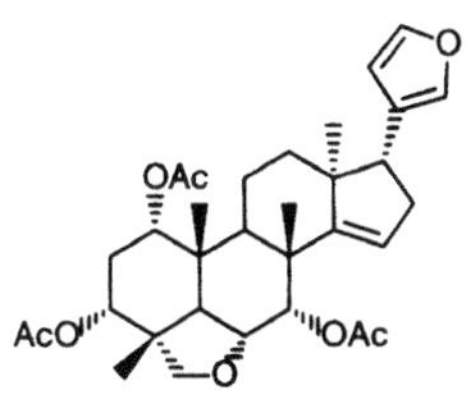 Vilasinin triacetate (**53**)	$C_{32}H_{42}O_8$; 554.2879 **m.p.** 228°C **Isolation**: Seed oil	*197*

Table 8 *(continued)*

Vilasinin group		Ref.
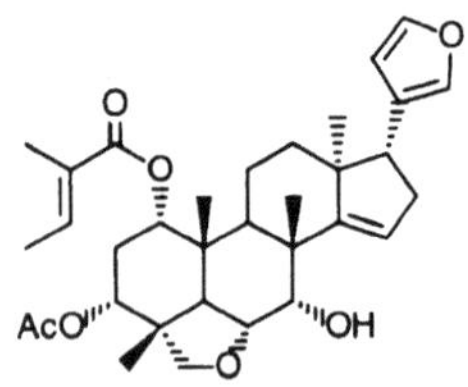 1-Tigloyl-3-acetylvilasinin (**54**)	$C_{33}H_{44}O_7$; 554.3087 **Isolation**: Seeds	*121*
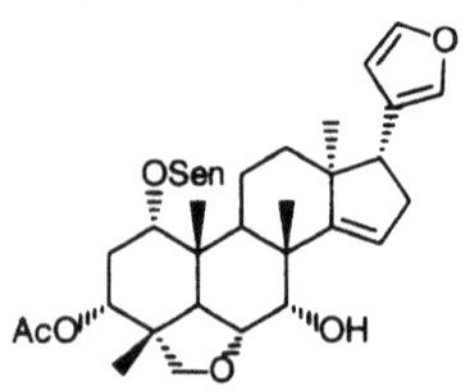 1-Senecioyl-3-acetylvilasinin (**55**)	$C_{33}H_{44}O_7$; 552.3087 **Isolation**: Leaf extract	*131*
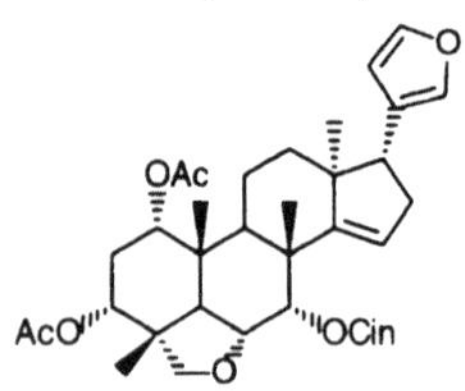 1,3,-Diacetyl-7-cinnamoylvilasinin or Nimbolin A (**56**)	$C_{39}H_{46}O_8$; 642 **m.p.** 180–183°C $[\alpha]_D$ −38.6° **Isolation**: Petroleum ether extract of ground wood **Spectra**: ^{1}H NMR	*71*
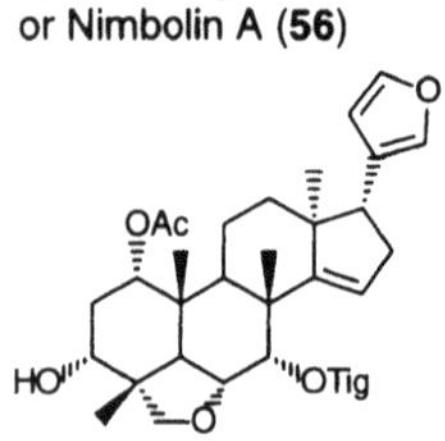 1-Acetyl-7-tigloylvilasinin (**57**)	$C_{33}H_{44}O_7$; 554.3087 **Isolation**: Seeds	*121*
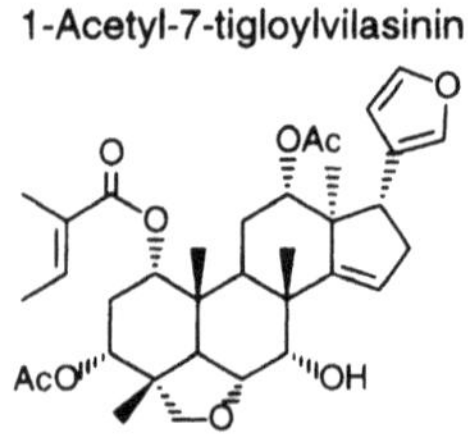 1-Tigloyl-3-acetyl-12α-acetoxy-vilasinin (**58**)	$C_{35}H_{46}O_9$; 610.3141 **Isolation**: Dried seeds	*125, 131*

Table 8 *(continued)*

Vilasinin group		Ref.
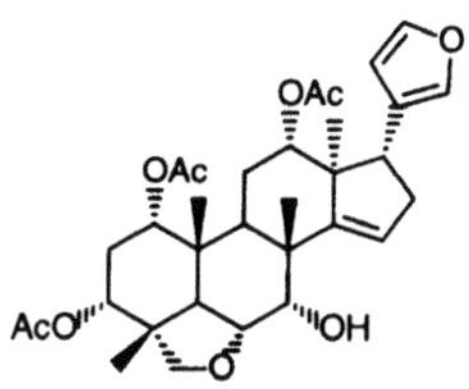 1,3-Diacetyl-12α-acetoxy-vilasinin (**59**)	$C_{32}H_{42}O_9$; 570.28282 **Isolation**: Dried seeds	*125, 131*
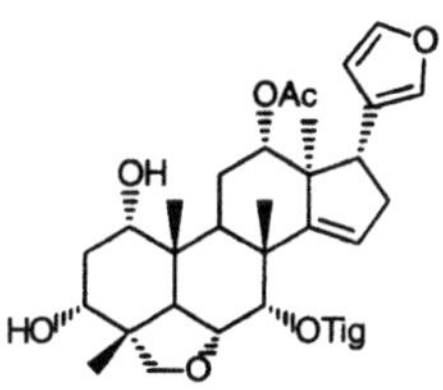 7-Tigloyl-12α-acetoxy-vilasinin (**60**)	$C_{33}H_{44}O_9$; 568.3036 **Isolation**: Dried seeds	*125, 131*
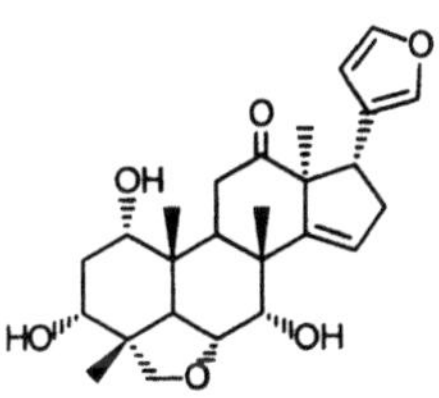 12-oxovilasinin or Nimbidinin (**61**)	$C_{26}H_{34}O_6$; 442 **m.p.** 282–84°C **Isolation**: neutral fraction of seed kernels **Spectra**: UV, IR **Derivative**: Triacetate, $C_{32}H_{40}O_9$, m.p. 222–24°, NMR	*173, 174*
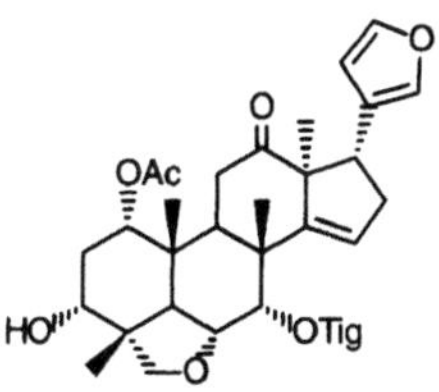 1-Acetyl-7-tigloylnimbidinin (**62**)	$C_{33}H_{42}O_8$; 566.2879 **Isolation**: Dried seeds	*125, 131*

Table 8 *(continued)*

Vilasinin group		Ref.
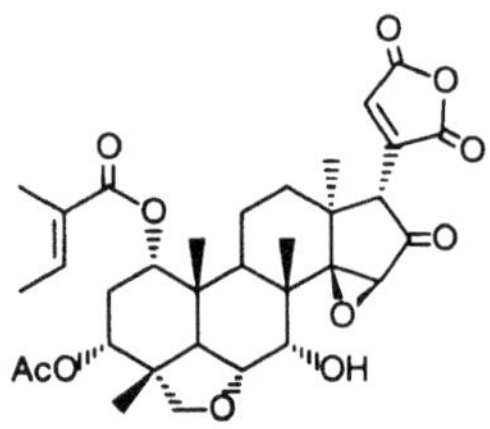 Limbocinin (**63**)	$C_{33}H_{40}O_{11}$; 612.2570 **Isolation**: EtOH extract of seeds	*261*
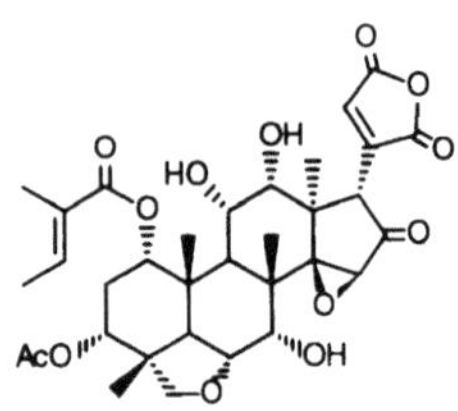 Limbocidin (**64**)	$C_{33}H_{40}O_{13}$; 644.2468 **Isolation**: EtOH extract of seeds	*261*
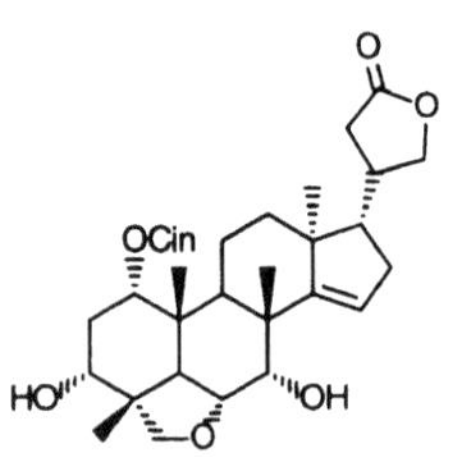 1-Cinnamoylvilasinin lactone (**65**)	$C_{35}H_{44}O_{7}$; 576.3087	*125*
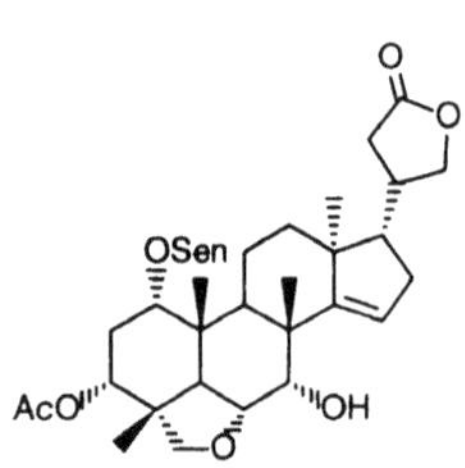 1-Senecioyl-3-acetylvilasinin lactone (**66**)	$C_{33}H_{46}O_{8}$; 570.3192 unpublished results (W. KRAUS and R. CRAMER)	

Table 8 *(continued)*

Vilasinin group		Ref.
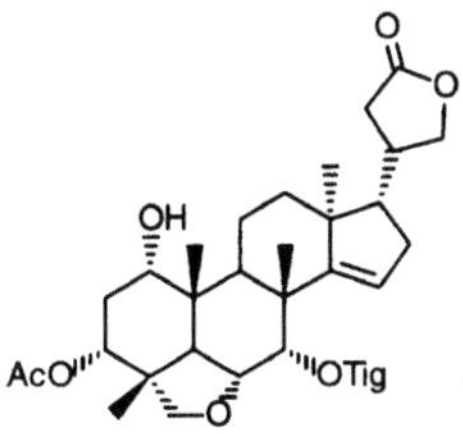 3-Acetyl-7-tigloylvilasinin lactone (**67**)	$C_{33}H_{46}O_8$; 570.3192 **m.p.** 242–43°C $[\alpha]_D^{20}$ −22.4° ($CHCl_3$, c 1.6) **Isolation**: HPLC of EtOH extract of neem seeds **Spectra**: IR, ^{1}H NMR, X-rays	*35*
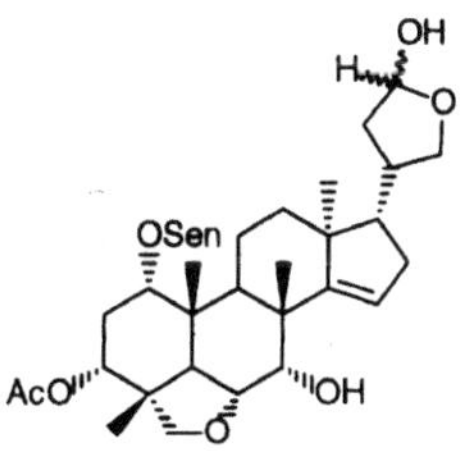 1-Senecioyl-3-acetylvilasinin lactol (**68**)	$C_{33}H_{48}O_8$; 572.3349 Unpublished results (W. Kraus and R. Cramer)	
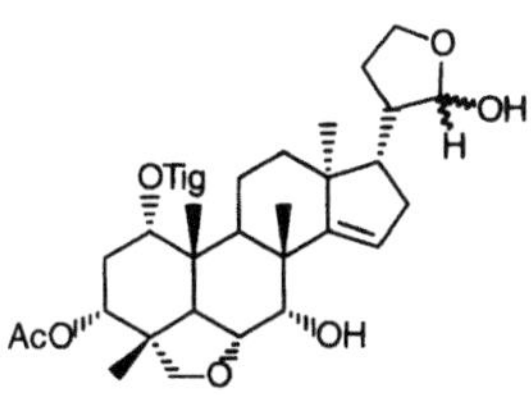 1-Tigloyl-3-acetylvilasinin lactol (**69**)	$C_{33}H_{48}O_8$; 572.3349	*125*
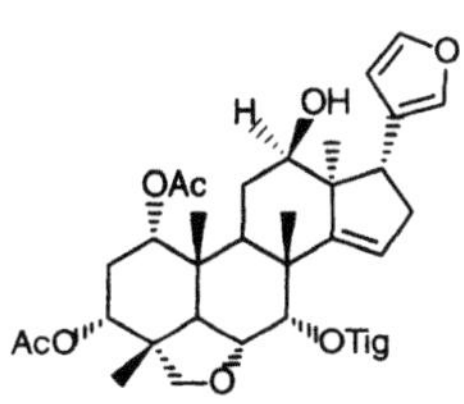 1,3-Diacetyl-7-tigloyl-12-hydroxyvilasinin (**70**)	$C_{35}H_{46}O_9$; **m.p.** 248–250°C $[\alpha]_D^{30}$ 10.8° ($CHCl_3$, c 1.0) $\mathbf{R_f}$ 0.465 (10% acetone-$CHCl_3$) **Isolation**: Oil obtained from $CHCl_3$ extract of cleaned, powdered seeds **Yield**: 0.025 g/10.0 kg seeds	*141*

Table 8 *(continued)*

Vilasinin group		Ref.
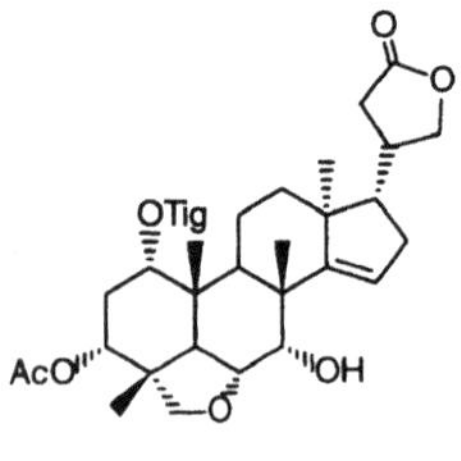 Azadirachtolide (**71**)	**Spectra**: IR, ^{1}H NMR, ^{13}C NMR, DEPT, COSY, FAB-MS $C_{33}H_{46}O_8$; 570.3192 **m.p.** 269°C $[\alpha]_D$ −48° ($CHCl_3$, c 4.25) **Isolation**: $CHCl_3$ extract of air-dried leaves **Yield**: 16.5 mg/200 g leaves **Spectra**: IR, ^{1}H NMR, ^{13}C NMR, NOESY, HMQC, HMBC, COSY	*208*
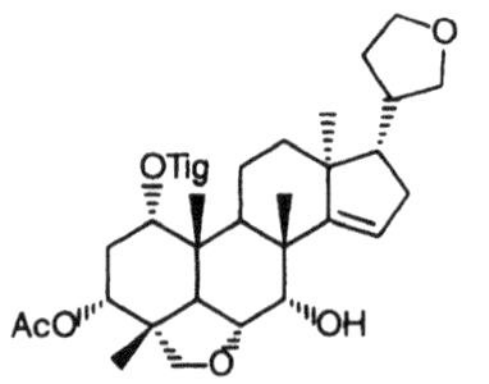 Deoxyazadirachtolide (**72**)	$C_{33}H_{48}O_7$; 556.34 **m.p.** 118–120°C **Isolation**: $CHCl_3$ extract of air-dried leaves **Yield**: 10.5 mg/200 g leaves **Spectra**: IR, ^{1}H NMR, ^{13}C NMR, NOESY, COSY, HMBC, HMQC	*208*
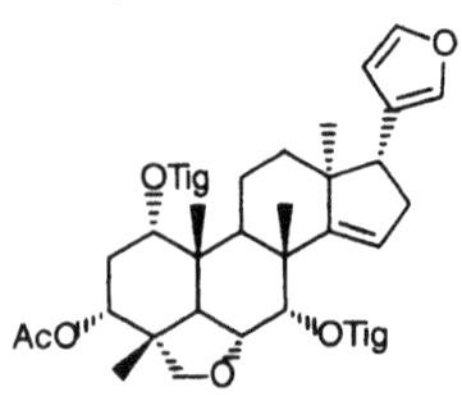 1,7-Ditigloyl-3-acetyl-vilasinin (**73**)	$C_{38}H_{50}O_8$; 634.3505 **m.p.** 180–82°C (colourless crystals) **Isolation**: EtOH extract of roots of *Melia composita* **Yield**: 400 mg/3 kg of dried roots **Spectra**: IR, ^{1}H NMR, ^{13}C NMR **Derivatives**: Hydrolysis: vilasinin	*288*
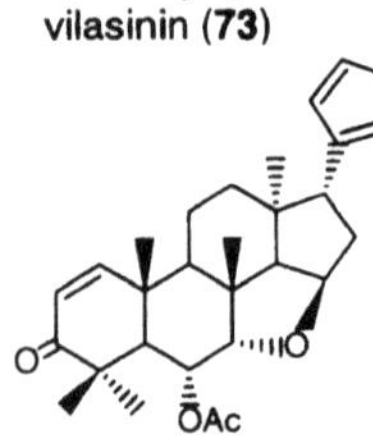 Vepinin (**74**)	$C_{28}H_{36}O_5$; 452 $[\alpha]_D$ 14.6° ($CHCl_3$) **Isolation**: Seed oil **Yield**: 0.15% **Spectra**: IR, ^{1}H NMR	*187*
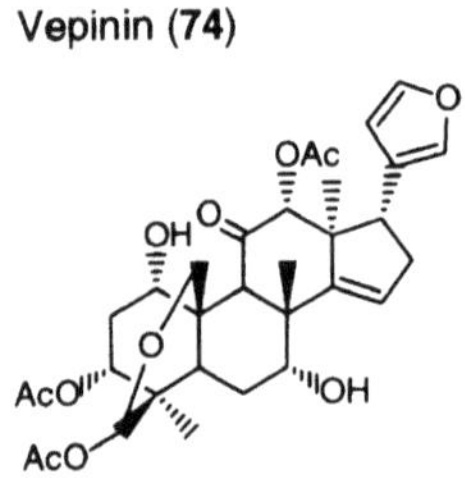 Azadirachtanin (**75**)	$C_{32}H_{40}O_{11}$; **m.p.** 225–226°C (colourless needles, E_2O-MeOH) $[\alpha]_D$ 8.7° ($CHCl_3$, c 0.45) **Isolation**: EtOH extract of fresh leaves **Yield**: 56 mg/3 kg leaves **Spectra**: UV, IR, ^{1}H NMR, ^{13}C NMR, MS **Derivative**: triketone, m.p. 205–207°C (flakes, Et_2O-MeOH), $[\alpha]_D$ 18.6° ($CHCl_3$, c 1.25), ^{1}H NMR	*205*

C-Seco Meliacins

This is a large group of compounds containing the most complex compounds found in Neem (*A. indica* or *Melia azadirach*). Broadly speaking, these compounds can be divided into three groups, *i.e.* the nimbin, salannin and azadirachtin groups. However depending upon slight variations in the basic skeletons we have divided the compounds into several more groups as mentioned in the introductory section.

2.9. Nimbin Group

Nimbin (**76**), the major bitter principle of neem, has been reported and characterized by chemical degradation (*172*). The presence of a butenolide side chain was confirmed by careful ozonolysis. Several chemical reactions were successfully carried out on this compound such as addition of two oxygen atoms to open the lactone double bond, sodium borohydride reduction, Oppenauer oxidation and acetylation to confirm its structure (*172, 248*). Extensive chemical modification studies were carried out and derivatives like nimbic acid, nimbinic acid (the partial methyl ester of nimbic acid), desacetylnimbin, dihydro- and hexahydronimbin etc. were prepared (*100*). NMR spectra showed the presence of a β-substituted furan ring (δ 7.34, 7.25, 6.35), two Carbomethoxy groups (δ 3.73, 3.64), one acetate (δ 3.03), a ketone group conjugated with a *cis* disubstituted double bond and a tertiary γ-carbon atom (typical AB quartet centred at δ 6.12, $J = 10$ cps), a cyclic ether and an isolated double bond (*184*). The stereochemistry was confirmed by double resonance NMR (*100*). The 1,3-diequatorial configuration of the ester groups on C-28 and C-6 was also suggested by the difference in the rate of hydrolysis of nimbin (**76**) and desacetylnimbin (**77**) and chemical evidence was provided for the stereochemistry at C-15 and C-17. ORD/CD measurements using pyronimbic acid established the absolute stereochemistry of the A/B ring fusion (*320*). The Confirmation of the structure assisted in establishing the structures of nimbin analogues.

6-Deacetylnimbinal (**81**) was the first C-4 tetranortriterpenoid aldehyde found in neem (*38, 125*). The presence of an aldehyde group was clear from an IR absorption at 2900 cm^{-1} and signals at δ 9.36 in the ^1H and δ 200.80 (d) in the ^{13}C NMR spectra. The methyls signals were assigned by means of NOE difference experiments and ^1H and ^{13}C COSY long range spectra. These techniques also showed that aldehyde group was attached at C-4. Deacetyl-2,3-dihydronimbic acid (**79**) has been synthesised from nimbolide (**87**) by partial hydrogenation and hydrolysis with dilute alkali (*70*). Nimbinal (**80**) (now called nimbanal) and nimbinol (**82**) have also been isolated and shown to be closely related to **81**. 6-Deacetylnimbinal (**81**) was converted to **80** by acetylation while sodium borohydride reduction produced nimbinol (**82**) (*38*). Various members of this group were shown to possess a γ-hydroxybutenolide side chain in place of the furan ring. 6-Deacetylnimbinolide (**83**), and 6-deacetylisonimbinolide (**85**) have been isolated from fresh stem twigs (*265*) whereas isonimbinolide (**84**) has been found in neem bark (*8*). Reaction of **76** with singlet oxygen produced **84** and **86** (*231*).

Table 9

Nimbin group		Ref.
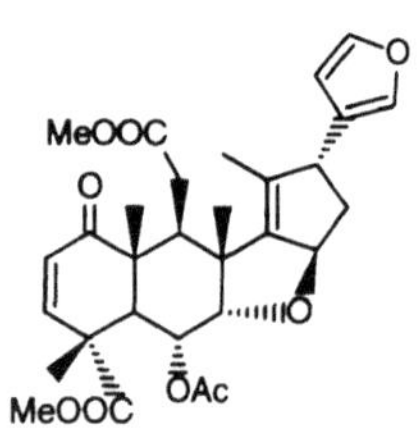 Nimbin (**76**)	$C_{30}H_{36}O_9$; 540	*65, 172, 185, 209, 248, 251, 312*
	m.p. 201–212°C	
	$[\alpha]_D^{20}$ **160°**	
	Isolation: (i) HPLC of extract of seed kernels	95
	(ii) From leaves and callus culture	*210*
	(iii) From Oil	*100*
	(iv) Neutral extract of petroleum ether extract of kernels	*275*
	Yield:(i) 210 mg/4 kg seed kernels	*95*
	(ii) 1.9 gm/1 kg seed oil	*100*
	Spectra: UV, IR, HRMS, ORD,	*275, 251, 320*
	negative ion MS, ^{1}H NMR,	*95, 168, 184, 227*
	Derivatives	*100, 185, 231*
	Biosynthesis: (i) Age factor in bark and callus	*236, 238*
	(ii) From glycine	*239*
	(iii) Biomimetic formation	*209*
	(iv) Callus tissue culture	*237*
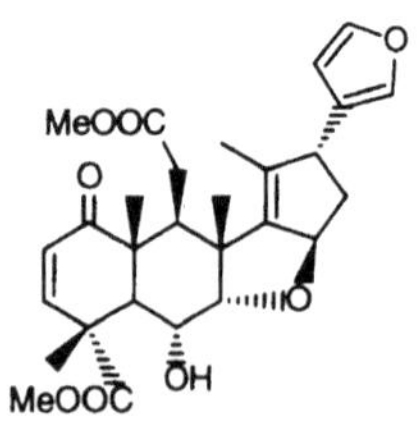 6-Deacetylnimbin (**77**)	$C_{28}H_{34}O_8$; 498.2252	*98, 168, 265, 275*
	m.p. 210–212°C (MeOH, Rods)[77], 208–210°C (fine needles)	*275*
	$[\alpha]_D$ 110°	
	Isolation: (i) Acidic fraction of CH_2Cl_2 extract of fresh, undried, uncrushed twigs	*265*
	(ii) HPLC	*95*
	(ii) Neutral fraction of pet. ether extract of seed kernels	*275*
	Yield: 200 mg/4 kg seed kernels	*95*
	Spectra: UV, IR, MS, negative ion MS	

Table 9 *(continued)*

Nimbin group		Ref.
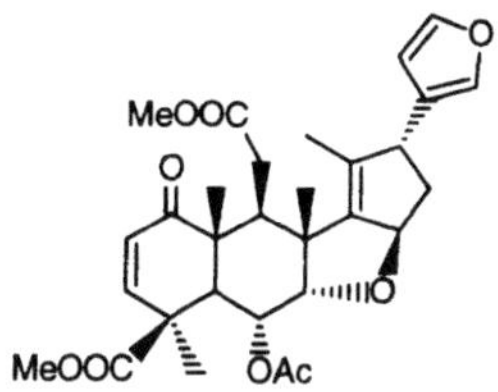 4-Epinimbin (**78**)	$C_{30}H_{36}O_9$; 540 **m.p.** 196–197°C (colourless needles, crystallised from MeOH) $[\alpha]^{27}_{D}$ 150° ($CHCl_3$, c 1.0) **Isolation**: Seed oil obtained from hexane extract of fresh seeds **Yield**: 2.5 g/1 kg seed oil **Spectra**: UV, IR, ^1H NMR, ^{13}C NMR, HRMS **Derivatives**: Hydrolysis: Epinimbic acid, m.p. 173°C (EtOAc-Hexane, colourless plates), IR	64
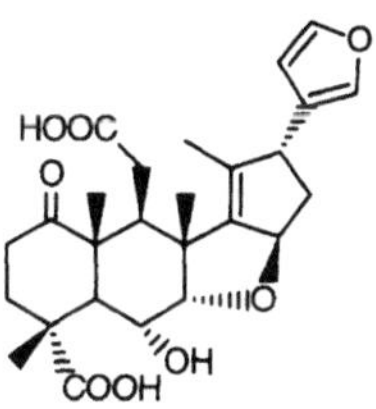 Deacetyl-2,3-Dihydro-nimbic acid (**79**)	$C_{26}H_{32}O_8$: 472.209 **m.p.** 192–194°C $[\alpha]_D$ 217° (pyridine)	*70*
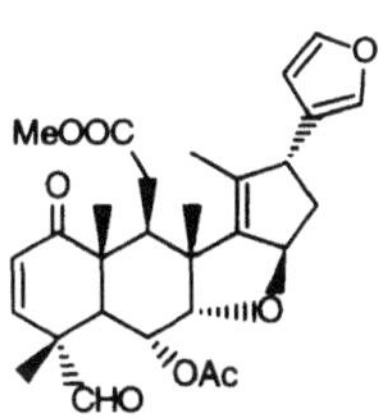 Nimbinal or Nimbanal (**80**)	$C_{29}H_{34}O_8$; 510 **m.p.** 195–197°C (colourless crystals, crystallised from Me_2CO-petrol) $[\alpha]_D$ 46° **Isolation**: EtOH extract of dried neem seeds **Yield**: 125 mg/0.9 kg (0.014%) **Spectra**: UV, IR, ^1H NMR, ^{13}C NMR, MS **Derivative**: Hydrogenation: 1,2-Dihydronimbanal, UV, IR, ^1H NMR, MS	*217*
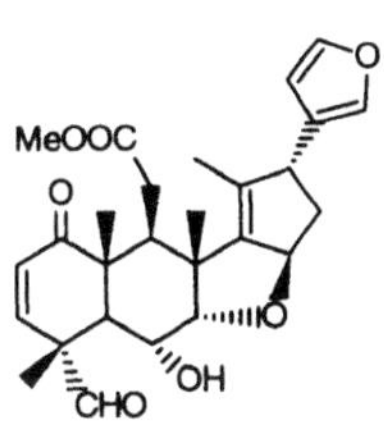 6-Deacetylnimbanal (**81**)	$C_{27}H_{32}O_7$; 468.2143 **m.p.** amorphous $[\alpha]^{\lambda}_{20^{\circ}C}\ \frac{589\quad 578\quad 546\quad 436\quad 405\quad 365}{+18\quad +20\quad +24\quad -56\quad -158\quad -664}$ ($CHCl_3$, c 0.1) **Isolation**: Diethyl ether extract of finely powdered air dried neem leaves **Yield**: 1.2 g/10 kg dried leaves **Spectra**: IR, ^1H NMR, ^{13}C NMR, COSY, NOE, EIMS **Derivatives**: (i) Acetylation: nimbinal (ii) Reduction of nimbinal: Nimbinol, amorphous, ^1H NMR, ^{13}C NMR, EIMS	*38*

Table 9 *(continued)*

Nimbin group		Ref.
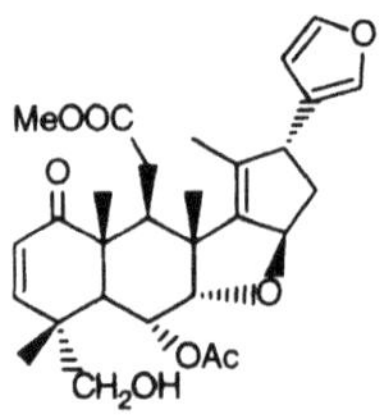 Nimbinol (**82**)	$C_{29}H_{36}O_8$; 512.2397 **m.p.** amorphous $[\alpha]^{\lambda}_{20^\circ C}$ $\frac{589 \quad 578 \quad 546 \quad 436 \quad 405 \quad 365}{+172 \quad +185 \quad +206 \quad +323 \quad +344 \quad -13}$ ($CHCl_3$, c 0.1) **Isolation**: MeOH extract of seed kernels **Yield**: 3 mg/3.5 kg seed kernels; 0.00046% of the extract **Spectra**: IR, ^1H NMR, ^{13}C NMR, NOE, EIMS	*38*
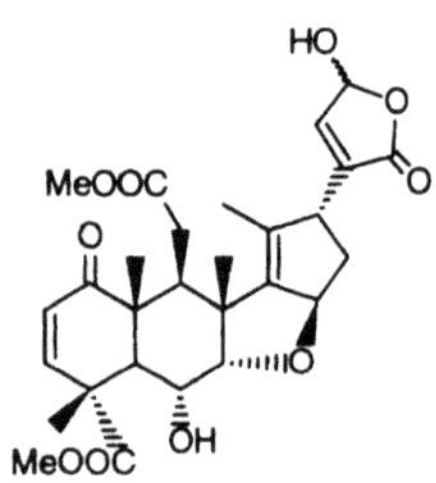 6-Deacetylnimbinolide (**83**)	$C_{28}H_{34}O_{10}$; 530.2150 **m.p.** 195–196°C (prismatic rods, crystallised from $CHCl_3$) $[\alpha]_D$ 100° ($CHCl_3$, c 0.02) **Isolation**: Acidic fraction of CH_2Cl_2 extract of fresh, undried, uncrushed twigs **Yield**: 10 mg/6 kg (0.0003% on dry wt. Basis) **Spectra**: UV, IR, ^1H NMR, ^{13}C NMR, NOESY, MS **Derivative**: Acetate, m.p. 130–132° (rods, $CHCl_3$), UV, IR, MS	*265*
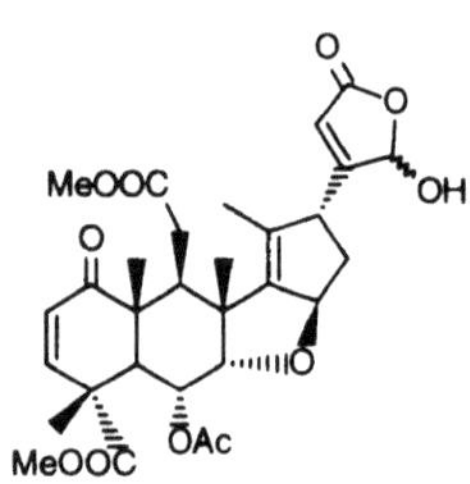 Isonimbinolide (**84**)	$C_{30}H_{36}O_{11}$; 572.228 **m.p.** 172–173°C (needles, crystallised from MeOH) **Isolation**: Neutral fraction of EtOH extract of stem bark **Spectra**: IR, ^1H NMR, COSY, NOESY, MS **Activity**: Insect growth regulating and antifeedant properties	*8*
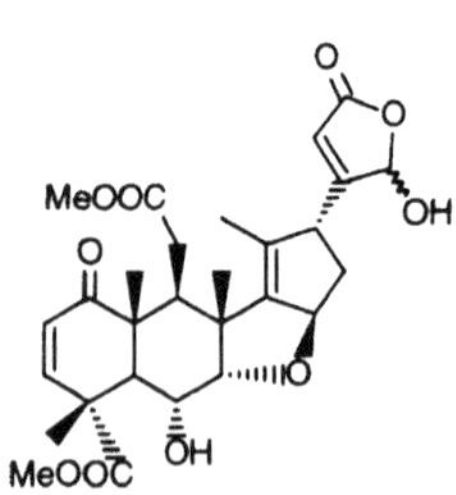 6-Deacetylisonimbinolide (**85**)	$C_{28}H_{34}O_{10}$; 530.2170 **m.p.** 178–180°C (bunches of rods, crystallised from MeOH) $[\alpha]_D$50° ($CHCl_3$, c 0.06) **Isolation**: Acidic fraction of CH_2Cl_2 extract of fresh, undried, uncrushed twigs **Yield**: 15 mg/6 kg (0.0004% on the dry wt. Basis) **Spectra**: UV, IR, ^1H NMR, ^{13}C NMR, NOESY, MS **Derivative**: Acetate, m.p. 112–114° (needles, $CHCl_3$), UV, IR, MS	*265*

Table 9 *(continued)*

Nimbin group		Ref.
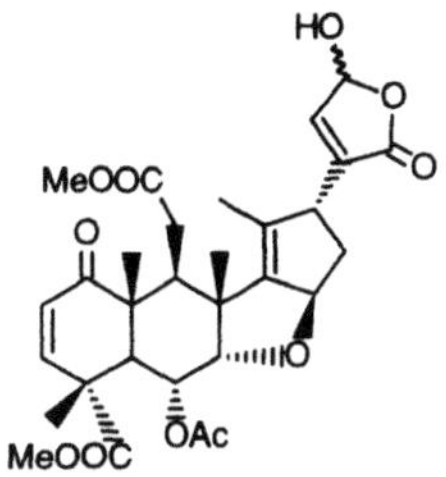 Nimbinolide (**86**)	$C_{30}H_{36}O_{11}$; 572.225	*124*

2.10. Nimbolide Group

Nimbolide (**87**) has been isolated from leaves (*70*). Its biosynthesis has been studied in detail by several groups using radioactive precursors (*72*, *73*, *75*). 28-Deoxonimbolide (**88**) has been reported in leaf extracts (*38*); its structure was established using ^{1}H and ^{13}C NMR data which showed that it was very similar to **87** (*113*) except for the absence of the lactone carbonyl signal (δ 175.00). Instead a triplet at δ 79.08 (t) characteristic of an oxymethylene carbon was identified as that of C-28 by NOE, ^{1}H, and ^{13}C spectroscopy. The crystalline acidic constituent nimbidic acid (**89**) has been found to be identical with salannic acid derived from salannin (**107**) (*174*). Its NMR spectrum revealed the presence of only one carbomethoxy methyl at δ 3.6 ppm showing it to be a monocarboxylic acid. Nimbidic acid (**89**) on acetylation (Py/Ac_2O) afforded a lactone acetate, 3-acetylnimbidic acid-1,12-lactone. This spontaneous tendency to lactonise is in agreement with the absence of the molecular ion peak ($M^+ = 482$) in the mass spectrum of nimbidic acid. Ochinin acetate (**90**) has been isolated from neem seed extract along with ochinolide B (**99**) (*231*).

Biosynthesis. The tetranortriterpenoids of this class are very important because of their biological activities. A number of hypotheses have been proposed for the opening of ring C and formation of the C-7α/15β ether bridge during biosynthesis of nimbin (**76**). It has been presumed to involve a hemiacetal of type **Y** (Scheme 6) (*295*) formed from a vilasinin-derived precursor (**T**), (12-hydroxyhavanensin) with an oxgyenated C-12, which may give rise to an aldehyde (**U**) further

Table 10

Nimbolide group		Ref.
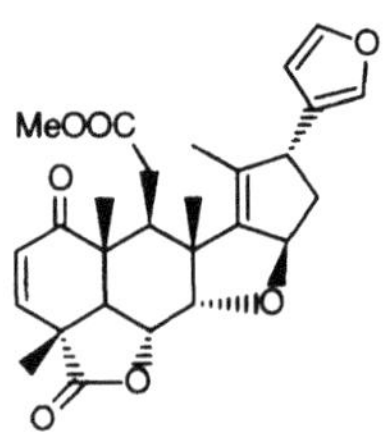 Nimbolide (**87**)	$C_{27}H_{30}O_7$; 466 **m.p.** 245–47°C[57], 228–230°C[93] $[\alpha]_D$ 206°, $[\alpha]_D^{23}$ 188°, $[\alpha]_D^{25}$ 211° ($CHCl_3$, c 0.1) **Isolation**: (i) HPLC (ii) $CHCl_3$ extract of leaves dipped in NH_3 and then dried (iii) Petroleum ether extract of fresh leaves (iv) $CHCl_3$ soluble fraction of EtOH extract of air-dried leaves **Yield**: 900 mg/4 kg neem kernels; 133 mg/10 kg leaves **Spectra**: UV, IR, ^1H NMR, ^{13}C NMR, INEPT, NOE, COSY, HETCOR, MS, negative ion MS **Derivatives**: Deacetyl-dihydronimbic acid, m.p. 192–194°C, $[\alpha]_D$ 217° (pyridine) **Biosynthesis**: (i) From [2-^{14}C]-acetate and mevalonate in leaves (*73, 75*) (ii) From [2-^{4}C, 4R-$^3H_1$]mevalonic acid in leaves (*72*) **Activity**: Anticancer (*113*), antimalarial (*225*), mutagenic and antibacterial (*226*), toxicity (*87*)	*70, 72, 75, 84, 95, 113, 168* *70, 84, 113*
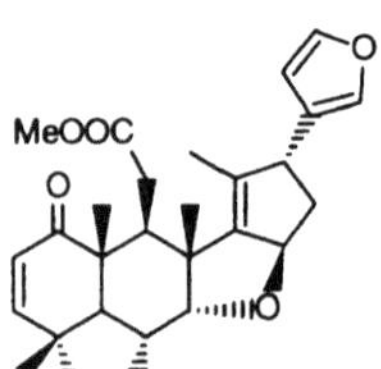 28-Deoxonimbolide (**88**)	$C_{27}H_{32}O_6$; 452.2179 **m.p.** 142–144°C (*113*), 170°C (*38*) $[\alpha]^{\lambda}_{20\,°C}$ 589 578 546 436 405 ($CHCl_3$ / +203 +213 +248 +480 +643 c 0.1) $[\alpha]_D^{25}$ 228° ($CHCl_3$, c 0.103) **Isolation**: (i) $CHCl_3$ extract of leaves dipped in NH_3 and then dried (*113*) (ii) Diethyl ether extract of dried leaves (*38*) **Yield**: 1.99 g/10 kg leaves **Spectra**: UV, IR, ^1H NMR, ^{13}C NMR, COSY, NOE, HETCOR, MS **Activity**: Anticancer (*113*)	*38, 113*
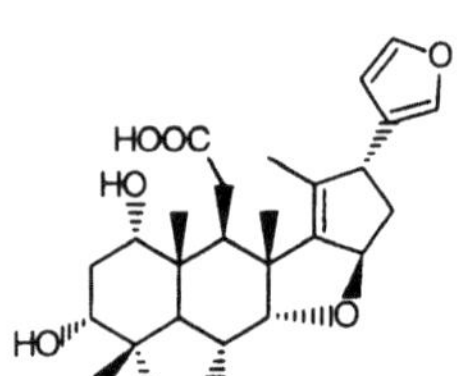 Nimbidic acid (**89**)	$C_{26}H_{34}O_7$; 440 **m.p.** 228–230°C **Isolation**: Acidic fraction of seed kernel extract **Spectra**: IR **Derivative**: Methyl ester, $C_{27}H_{36}O_7$, m.p. 221–23°C, NMR	*173, 174*

Table 10 *(continued)*

Nimbolide group		Ref.
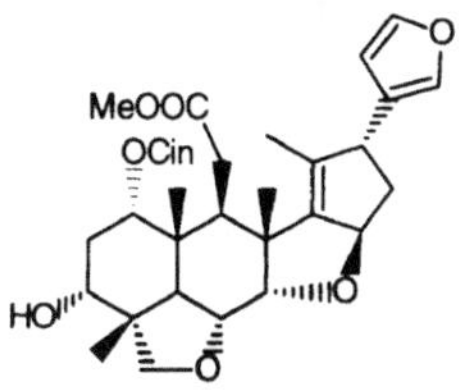 Ochinin acetate (**90**)	$C_{36}H_{42}O_8$; 602.2879	*193, 231*

hydrated to an intermediate aldehyde hydrate (**V**). The latter reaction is important because it may initiate rotation around C-8/C-14 followed by inversion of configuration at C-15, thus generating hemiacetal (**Y**). Oxidation of (**Y**) would produce the ohchinolide-type compounds (**99**, **100**, **105**, **106**) and hydrolysis followed by esterification would produce nimbolidins (**101**, **102**, **103**, **104**) both in Table 12. However, it seems more likely that the nimbins are formed from 12-hydroxynimocinol-type or sendanal-type precursors (**w**, Scheme 6) possibly via the fragmentation/cyclisation reaction sequence shown in the Scheme. Intermediates like **w**′, **w**″ and **z** may also be involved in the biosynthesis of nimbolide (**87**) and 28-deoxonimbolide (**88**). Several biosynthetic studies using [^{3}H]euphol, [^{3}H]butyrospermol, [^{3}H]tirucallol and [^{3}H]-Δ^7-tirucallol (*73*), [2-^{14}C]mevalonate (*72*, *73*), [2-^{14}C]acetate (*72*) and [2-^{14}C, (^4R)4-$^3H_1$] mevalonate (*65*) have been carried out. It has been concluded that the biosynthesis of nimbolide from mevalonate occurs in the leaves, that euphol is more efficiently incorporated than tirucallol and that Δ^8-isomers (**D**) are more effectively utilised than the Δ^7-isomers (**F**) (Scheme 2). These results further suggested that the biosynthesis of nimbolide (**87**) may involve a $\Delta^{7,9(11)}$-diene formed from an 8α,9α-epoxide (*72*, *73*, *75*). The 7α-hydroxy-apo-systems would then be derived from the Δ^7-function of the diene, whereas the $\Delta^{9(11)}$-function would activate the C-12 position for oxidation followed by fragmentation to give the C-seco compounds.

2.11. Nimbinene Group

Pentanortriterpenoids have been isolated for the first time from different plant parts. Nimbinene (**91**) and 6-deacetylnimbinene (**92**) have

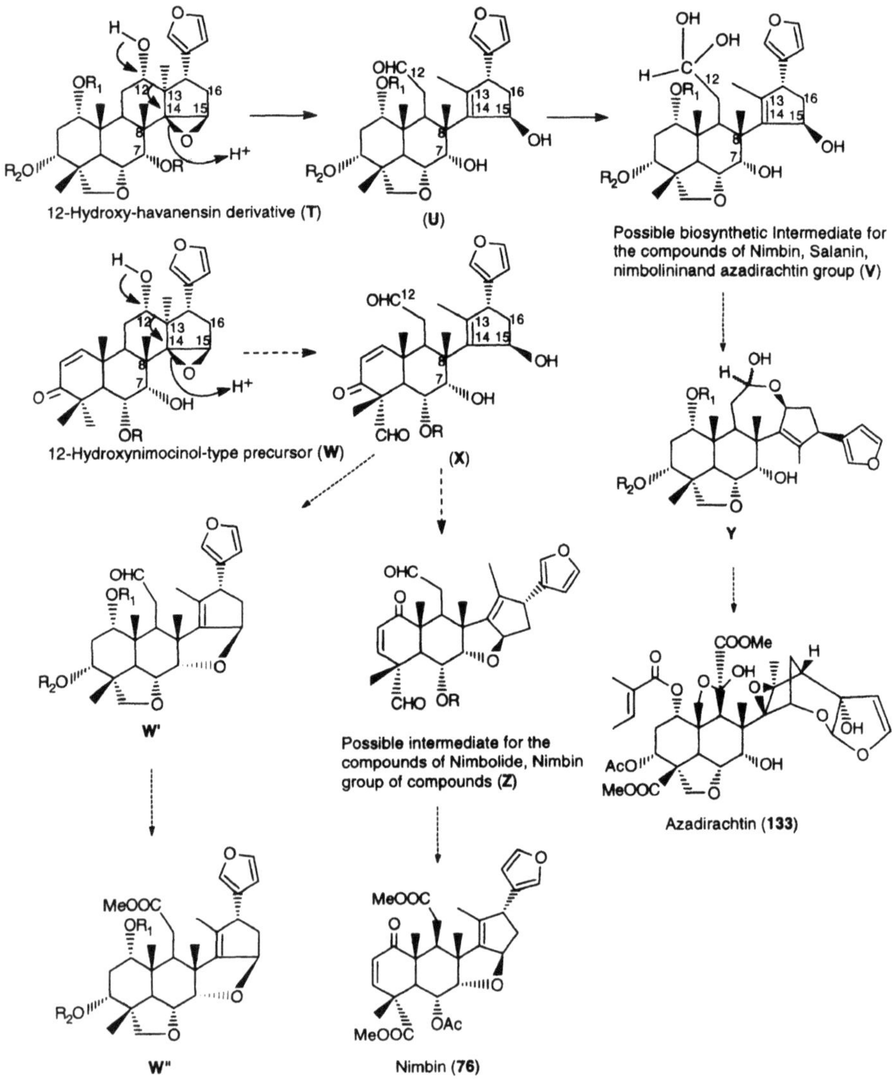

Scheme 6. A possible mechanism for the cleavage of ring C and biosynthesis of compound having C-seco skeleton; major effective limonoids of neem fall in this category

been isolated from seed oil, leaves and bark, nimbandiol (**93**) form seed oil and leaves and 6-*O*-acetylnimbandiol (**94**) only form seed oil (*135*). Structures of all these compounds were established by ^{1}H and ^{13}C NMR spectroscopy (*135*, *139*). 4α-Benzoylnimbandiol (**95**) has been isolated from the methanol extract of seeds along with **93**. Its ^{1}H NMR spectrum

Table 11

Nimbinene group		Ref.
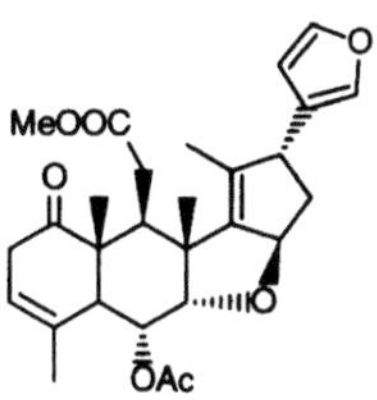 Nimbinene (**91**)	$C_{28}H_{34}O_7$; 482.6 **m.p.** 134°C (MeOH) $[\alpha]_D^{20}$ 168° ($CHCl_3$, c 1.0) **Isolation**: (i) Seed oil obtained by petroleum ether – MeOH (50 : 50) extraction (ii) Ether extract of leaves and bark **Yield**: 350 mg/10 kg neem extractive (547 g oil); 250 mg/10 kg leaves (283 g extract); 6 g/20 kg bark (270 g extract) **Spectra**: IR, ^{1}H NMR, ^{13}C NMR, MS	*135*
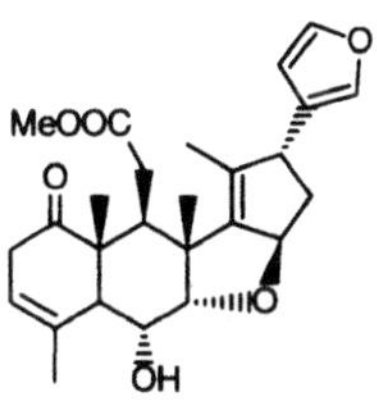 6-Desacetylnimbinene (**92**)	$C_{26}H_{32}O_6$; 440.6 **m.p.** 141°C (MeOH) $[\alpha]_D^{20}$ 132° ($CHCl_3$, c 1.0) **Isolation**: (i) seed oil obtained by petroleum ether – MeOH (50 : 50) extraction (ii) Ether extract of leaves and bark **Yield**: 520 mg/10 kg extractive (547 g oil); 600 mg/10 kg leaves (283 g extract); 750 mg/20 kg bark (270 g extract) **Spectra**: IR ^{1}H NMR, ^{13}C NMR, MS **Derivative**: Acetate, m.p. 133°C	*135*
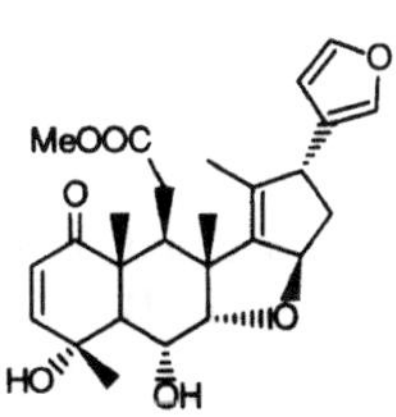 Nimbandiol (**93**)	$C_{26}H_{32}O_7$; 456.5 **m.p.** 121°C (ethyl acetate) $[\alpha]_D^{20}$ 187.9° ($CHCl_3$, c 1.0) **Isolation**: (i) seed oil obtained by petroleum ether – MeOH (50:50) extraction (ii) Ether extract of leaves **Yield**: 2.5 g/10 kg extractive (547 g oil); 1.25 g/10 kg leaves (253 g extract) **Spectra**: IR, ^{1}H NMR, ^{13}C NMR, MS **Derivative**: Acetate, m.p. 231°C, $[\alpha]_D^{20}$ 115.8° ($CHCl_3$, c 1.0)	*135*
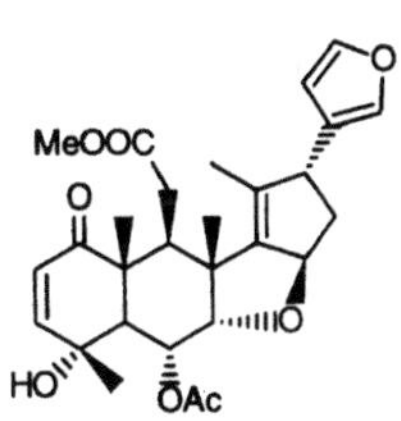 6-*O*-Acetylnimbandiol (**94**)	$C_{28}H_{34}O_8$; 498.6 **m.p.** 178°C (MeOH) $[\alpha]_D^{24}$ 232.5° ($CHCl_3$, c 0.03) (*139*) 245° ($CHCl_3$, c 1.0) (*135*) **Isolation**: (i) MeOH soluble fraction of oil of fresh fruits (ii) Seed oil obtained from petroleum ether – MeOH (50 : 50) **Yield**: 28 mg/300 g oil (*139*), 1.2 g/10 kg neem extractive (547 g seed oil) (*135*)	*135*, *139*

Table 11 *(continued)*

Nimbinene group		Ref.
	Spectra: IR, ^{1}H NMR, ^{13}C NMR, SI-MS **Activity**: Insect growth inhibitor and insecticide	
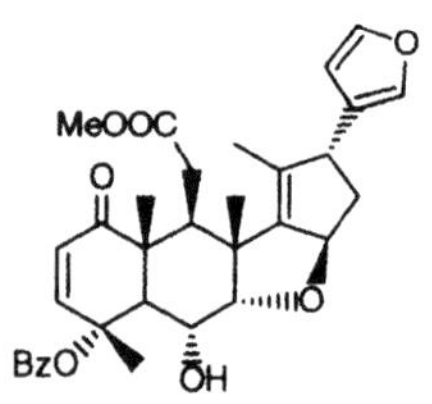 **4α-benzoylnimbandiol (95)**	$C_{33}H_{36}O_8$; 560.241 **Spectra**: ^{1}H NMR	*231*

was similar to that of **93** except for the signals due to hydroxyl groups. It did not show the signal at δ 4.29 due to the C-4 but displayed a multiplet at δ 7.48–7.82 assignable to a benzoate at this position at C-4. The downfield chemical shift of H-6, δ 4.27 (ddd, $J=10$, 7.5, 3.5 Hz) compared with δ 3.28 (d, $J=7.5$ Hz) in nimbandiol, was ascribed to the presence of a benzoyl group at C-4.

2.12. Nimbolinin Group

The petroleum ether extract of ground neem wood yielded 1,3-diacetyl-7-cinnamoylnimbolinin commonly known as nimbolin B (**96**) (*71*). Its ^{1}H NMR spectrum compared with that of nimbolin A (**56**) which was also reported from ground wood, exhibited two significant differences. While that of nimbolin A showed four tertiary methyls, the spectrum of nimbolin B showed only three, the fourth being replaced by a sharp downfield 3 proton singlet at δ 1.8 ascribed to a vinylic methyl not broadened by homoallylic coupling. Secondly, the vinyl 15-H signal of **56** is absent in **96** and in its place there is a 2H multiplet further upfield at δ 5.18. 1-Tigloyl-3-acetyl-7-cinnamoylnimbolinin or nimbilin (**97**) was isolated from the neutral fraction of the methylene chloride extract of root bark (*13*). The structure was determined by ^{1}H and ^{13}C NMR spectroscopy, the assignments being corroborated by a COSY-45 spectrum which showed connectivities of H-6 with H-5 and H-7, both H-1 and H-3 with H-2α and H-2β, H-15 with H-16α and H-16β, H-12 with H-11α and H-11β, H-2′ with H-3′, and H-21 with H-23, while the

Table 12

Nimbolinin group		Ref.
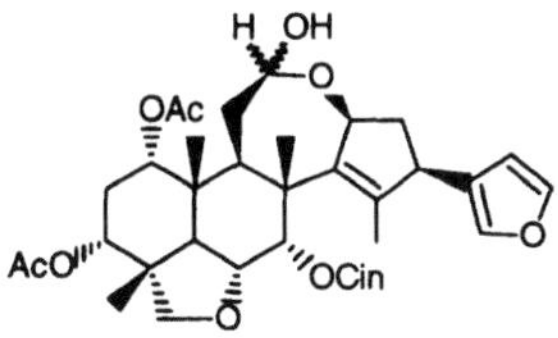 1,3-Diacetyl-7-cinnamoyl-nimbolinin (Nimbolin B) (**96**)	$C_{39}H_{46}O_{10}$; 674 **m.p.** 140–142°C (white crystalline powder) (*15*) 243–245°C (*71*) $[\alpha]_D$ −93.3° **Isolation**: (i) Petroleum ether extract of ground wood (ii) Neutral fraction of CH_2Cl_2 extract of root bark **Spectra**: IR, ^1H NMR, ^{13}C NMR, NOESY, COSY, HET-COSY, EIMS **Derivative**: Acetate, m.p. 162–168°C (irregular plates, petroleum ether), UV, IR, EIMS	*15*, *71* *71* *15*
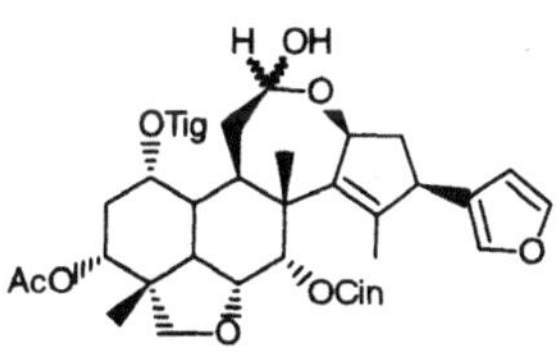 1-Tigloyl-3-acetyl-7-cinnamoylnimbolinin (Nimbilin) (**97**)	$C_{42}H_{50}O_{10}$; 714 **m.p.** 149–150°C (plates, crystallised from hexane) **Isolation**: Neutral fraction of CH_2Cl_2 extract of defatted root bark **Yield**: 11.9 g/20 kg root bark (0.047% of the dry wt. of neutral fraction) **Spectra**: UV, IR, ^1H NMR, ^{13}C NMR, MS **Derivative**: Acetate, m.p. 171–172°C (irregular plates, $CHCl_3$), UV, IR, MS	*13*
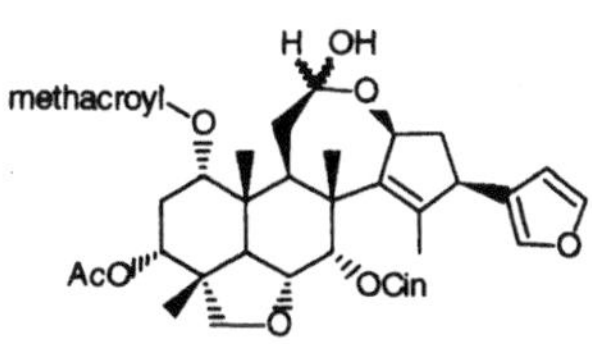 1-Methacroyl-3-acetyl-7-cinnamoylnimbolinin (Nimbolicin) (**98**)	$C_{41}H_{48}O_{10}$; 700 **m.p.** 121–122°C (irregular plates) $[\alpha]_D^{24}$ −33.3° ($CHCl_3$) **Isolation**: Neutral fraction of CH_2Cl_2 extract of root bark **Yield**: 16.3 mg/28 kg root bark **Spectra**: UV, IR, ^1H NMR, ^{13}C NMR, NOESY, COSY, HETCOR, EIMS **Derivative**: Acetate	*15*
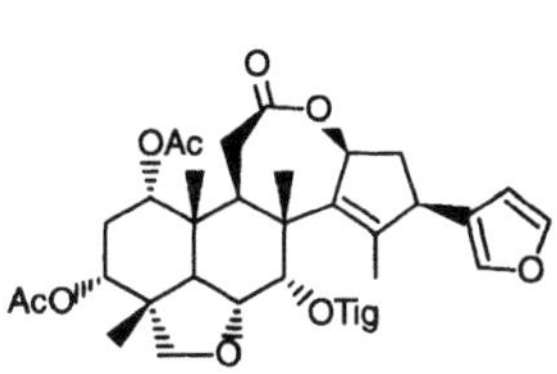 Ohchinolide B (**99**)	$C_{35}H_{44}O_{10}$; 624.2934 **m.p.** 210–212°C (MeOH) $[\alpha]_D^{20}$ −46.5° ($CHCl_3$, c 1.0) **Isolation**: (i) Ether extract of fruits (ii) HPLC **Yield**: 170 mg/5 kg fruits **Spectra**: IR, ^1H NMR, ^{13}C NMR **Derivative**: Nimbolidine B	*127*

Table 12 *(continued0*

Nimbolinin group		Ref.
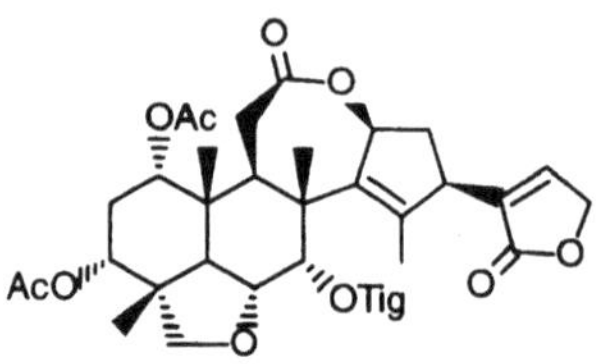 21-Oxo-ohchinolide (**100**)	$C_{35}H_{44}O_{11}$; 640.2883	*127*
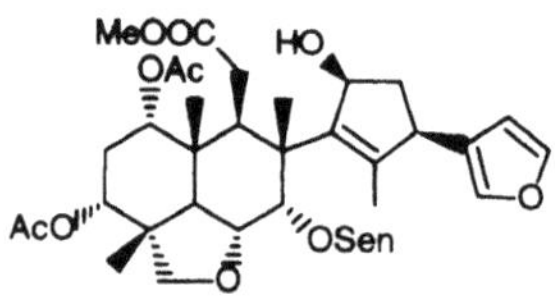 7-Detigloyl-7-Senecioyl-11-deacetylnimbolidin A (**101**)	$C_{36}H_{48}O_{11}$; 656.3196 **Isolation**: Leaf extract*	*124*
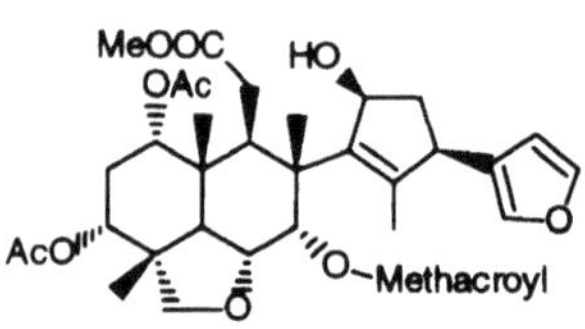 7-Detigloyl-7-methacroyl-11-deacetylnimbolidin A (**102**)	**Isolation**: Leaf extract*	*124*
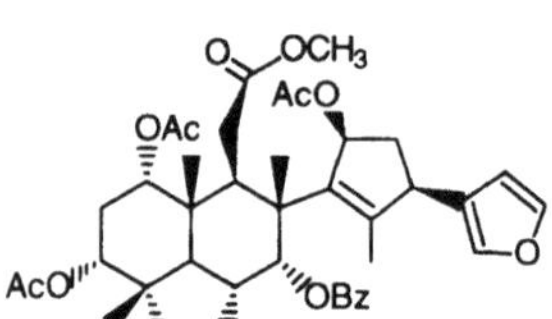 Nimbolidin A (**103**)	$C_{40}H_{48}O_{12}$; 720 **m.p.** 178°C (MeOH) $[\alpha]_D^{20}$ −32° ($CHCl_3$, c 1.0) **Isolation**: Petroleum ether and ether extract of fruits *Melia azadarach* Linn. **Yield**: 160 mg/5 kg fruits **Spectra**: IR, ^1H NMR, ^{13}C NMR, MS	*127*
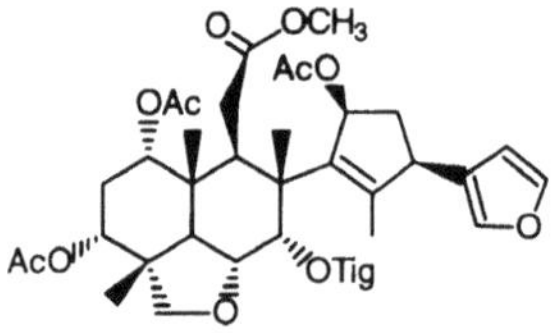 Nimbolidin B (**104**)	$C_{38}H_{50}O_{12}$; 698 **m.p.** 180°C (MeOH) $[\alpha]_D^{20}$ −9.2 ($CHCl_3$, c 1.0) **Isolation**: Petroleum ether and ether extract of fruits *Melia azadarach* Linn. **Yield**: 280 mg/5 kg fruits **Spectra**: IR, ^1H NMR, ^{13}C NMR, MS	*127*

Table 12 *(continued)*

Nimbolinin group		Ref.
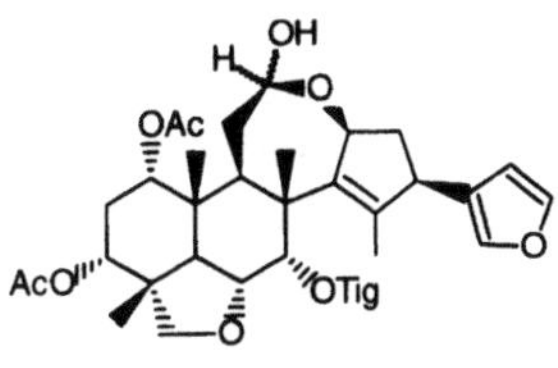 **Nimbolinin B (105)**	$C_{35}H_{46}O_{10}$; 626.309 $[\alpha]_D^{20}$ −55.5° ($CHCl_3$, c 0.91) **Isolation**: Petroleum ether and ether extract of fruits *Melia azadarach* Linn. **Yield**: 80 mg/5 kg fruits **Spectra**: IR, ^{1}H NMR, ^{13}C NMR, MS	*127*
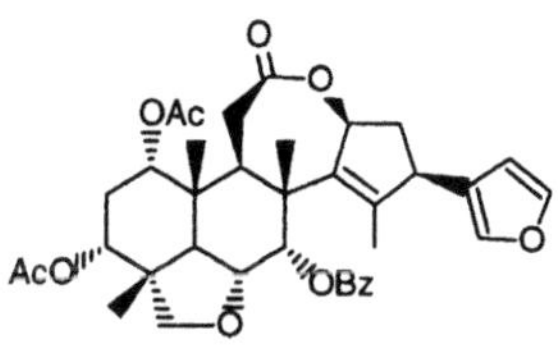 **Ohchinolide A (106)**	$C_{37}H_{42}O_{10}$; 646.2777 **m.p.** 230–31°C (MeOH) $[\alpha]_D^{20}$ −46.5° ($CHCl_3$, c 1.0) **Isolation**: Petroleum ether and ether extract of fruits of *Melia azadarach* Linn. **Yield**: 160 mg/5 kg fruits **Spectra**: IR, ^{1}H NMR, ^{13}C NMR, MS	*127*

* Original authors have put the names of compounds **101** and **102** as mentioned above. However, **101** and **102** should be named as 7-Detigloyl-7-senecioyl-15-deacetyl-nimbolidin-B and 7-Detigloyl-7-methacroyl-15-deacetyl nimbolidin-B, respectively.

stereochemistry of the various centres of nimbilin was established by the NOESY method. Nimbolicin (**98**) is another compound isolated from root barks whose structure was established similarly (*15*).

2.13. Salannin Group

Salannin (**107**) and its derivatives are characterized by the two oxygen bridges C-6 to 28 and C-7 to 14. **107** was the first compound of this group isolated from neem seeds and was subjected to extensive structural studies, spectroscopic investigations and chemical modifications (*101*). 3-Desacetylsalannin (**108**) was first prepared by selective hydrolysis of **107** and was later on isolated from neem seeds along with salannol (**109**) (*134*) and salannol acetate (**110**) (*121, 127*). Salannin (**107**) is second only to azadirachtin (**135**) in the attention given to it by chemists and several analytical methods have been developed to quantify the salannin content in neem extracts (*95, 318*). γ-Hydroxybutenolides are also known in the salannin group of compounds, with salannolide (**112**) being the first in the series to be isolated from neem seeds (*85*). 2′,3′-Dehydrosalannol (**111**) was isolated from $CHCl_3$ soluble fraction of dried leaves (*86*). Isoazadirolide (**113**), another γ-hydroxybutenolide, has been isolated from the acidic fraction of the fresh undried winter leaves (*272*). The ^{1}H NMR spectrum of **113** had signals at δ 5.06 (m, $W_{1/2} = 5.3$ Hz) and 4.91 (m, $W_{1/2} = 5.7$ Hz) ascribable to H-1 and H-3 respectively, five one proton multiplets at δ 5.50, 2.32, 2.22, 2.10 and 2.06 due to H-15, H-11, H-2β, H-16β and H-2α respectively. Five one proton doublets and four one proton double doublets have also been assigned. The stereochemistry of various centres of isoazadirolide has been established by the NOESY method which showed the spatial connectivities of H-17 with H-30, H-16β and H-7; H-30 with H-18, H-16β, H-11β, H-7, H-6 and H-1; H-19 with H-30, H-6 and H-2β; H-29 with H-28β and H-7; H-5 with H-28α and H-15; and also of H-6 with H-7. The spatial connectivity of H-17 with H-30, H-16β and H-7 indicated that the side chain at C-17 was α-oriented, while the spatial connectivity of H-30 with H-1 further confirmed the α-orientation of the ester substituent at C-1. Margosinolide (**114**) and isomargosinolide (**115**) have been isolated from fresh twigs (*255*); their structures are interesting because they uniquely incorporate both the ether linkage between C-28 and C-6 and the 1-en-3-one system of ring A. The compounds discussed so far which contain the ether linkage lack the 1-en-3-one system and instead possess oxygen substituents at C-1 and C-3 (*68*, *47*, *275*). Since the hydroxybutenolide side chain of **114** and **115** may be regarded as the intermediate in the formation of the furan ring it has been suggested (*255*) that **114** and **115** may be precursors of salannin (**107**) and that the formation of the latter's ring-A can be rationalised by invoking hydration of the double bond between C-1 and C-2 followed by reduction of the ketone function and esterification. Salannolactame-21 (**116**) and salannolactame-23 (**117**) are the only nitrogen-containing limonoids isolated so far (*138*).

Table 13

Salannin group		Ref.
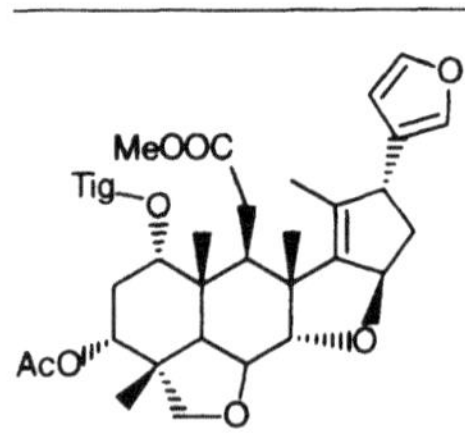 Salannin (**107**)	$C_{34}H_{44}O_9$; 596 **m.p.** 167–175°C; $[\alpha]_D$ 154° **Isolation**: (i) MeOH soluble fraction of neem oil of fresh fruits (ii) HPLC **Yield**: 900 mg/4 kg fruits **Spectra**: IR, ^{1}H NMR	*95*, *101*, *139*, *209*, *231*, *318* *318*, *95*

Table 13 *(continued)*

Salannin group		Ref.
	Derivatives: (i) Hydroxytigloylsalannin, m.p. 213–215°C (EtOAc-light petroleum), [α]$_D$ 137° (c 0.94), IR (ii) Salannin diol, m.p. 201–205°C (EtOAc–light petroleum), [α]$_D$ 135° (c 1.13), IR (iii) Destigloysalannin diacetate, m.p. 230–234°C ((EtOAc–light petroleum), [α]$_D$ 126° (c 1.1), IR (iv) Oxidation of diol (ii): ketol, m.p. 253–255°C (EtOAc), [α]$_D$ 160° (c 0.5), IR (v) Photooxidation: (a) Salannolide, $C_{34}H_{44}O_{11}$, m.p. 310° (decomp.) (b) Isosalannolide, ^{1}H NMR, ^{13}C NMR **Biosynthesis**: Biomimetic formation	*209*
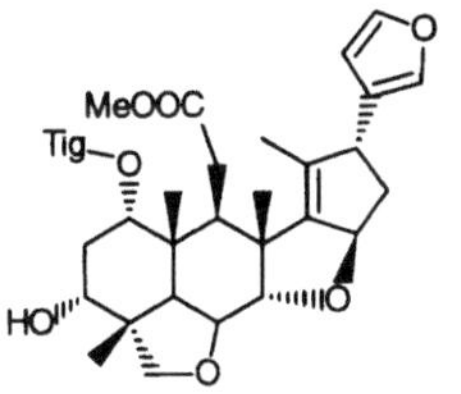 3-Deacetylsalannin (**108**)	$C_{32}H_{42}O_8$; 554.278 **m.p.** 195–200°C [α]$_D$ 54.2° ($CHCl_3$, c 0.05) **Isolation**: (i) $CHCl_3$ soluble fraction of EtOH extract of air-dried leaves (ii) MeOH soluble fraction of neem oil of fresh fruits **Yield**: 55 mg/300 g oil **Spectra**: IR, ^{1}H NMR, ^{13}C NMR, EIMS	*84, 134, 139*
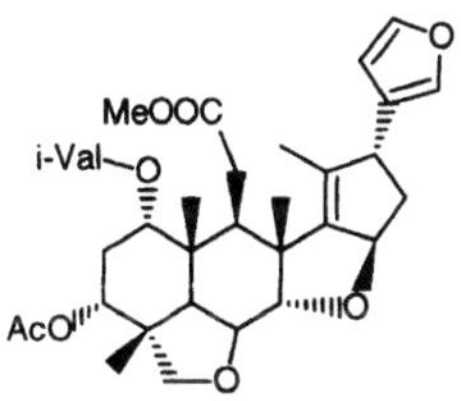 Salannol (**109**)	$C_{32}H_{44}O_8$; 556.3024 **m.p.** 208°C (EtOAc) [α]$^{20}_D$ 108.7° ($CHCl_3$, c 1.0) **Isolation**: Seed oil **Spectra**: IR, ^{1}H NMR, ^{13}C NMR, NOE	*134*
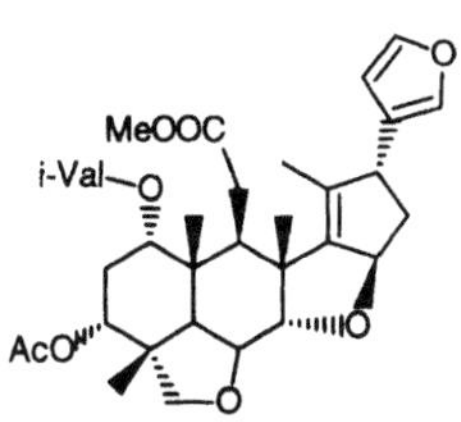 Salannol acetate (**110**)	$C_{34}H_{36}O_9$; 598 **m.p.** gummy mass [α]$_D$ 118° **Isolation**: EtOH extract of dried neem seeds **Yield**: 200 mg/0.9 kg seeds (0.022%) **Spectra**: IR, ^{1}H NMR, ^{13}C NMR, MS **Derivative**: Detigloylsalannol, m.p. 202–204°C (Me_2CO–petrol), IR, ^{1}H NMR, MS	227

Table 13 *(continued)*

Salannin group		Ref.
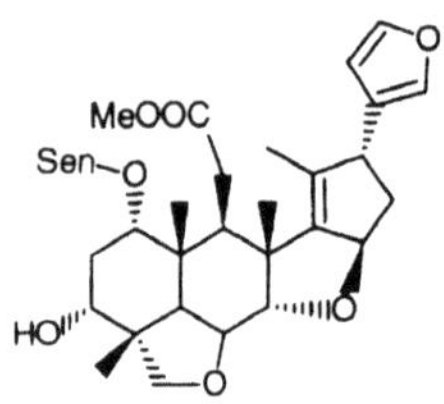 2',3'-Dehydrosalannol (**111**)	$C_{32}H_{42}O_8$; 554 **m.p.** 183–185°C (needles, MeOH – Et_2O) $[\alpha]_D$ 180° **Isolation**: $CHCl_3$ soluble fraction of EtOH extract of dried leaves **Yield**: 255 mg/8 kg leaves **Spectra**: IR, ^{1}H NMR, ^{13}C NMR, MS **Derivative**: Acetate, IR, MS	*86*
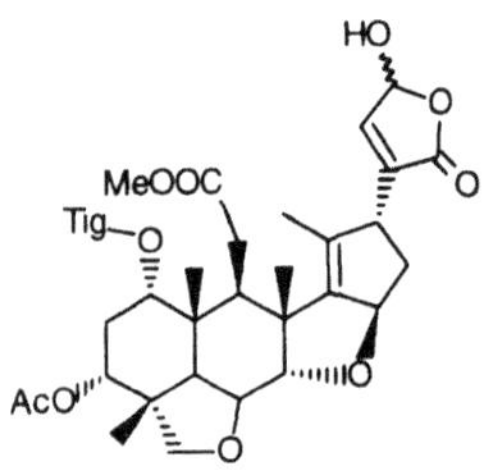 Salannolide (**112**)	$C_{34}H_{44}O_{11}$; 628 **m.p.** >320° (decomp.) $[\alpha]_D^{27}$ 185° ($CHCl_3$, c 1.0) **Isolation**: Column chromatography of bitter principles from fresh seed oil **Spectra**: IR, ^{1}H NMR, ^{13}C NMR, MS	*85*
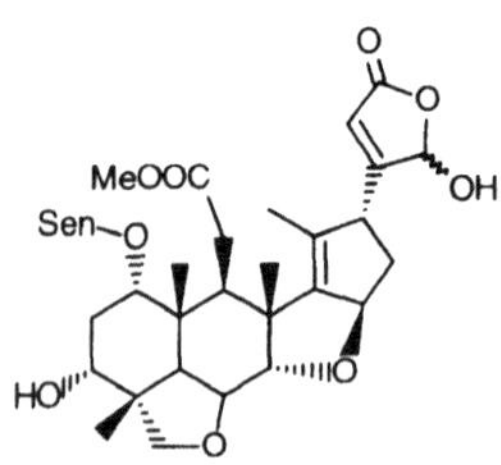 Isoazadirolide (**113**)	$C_{32}H_{42}O_{10}$; 586.2738 **m.p.** 130°C (rods) $[\alpha]_D^{24}$ 200° ($CHCl_3$) **Isolation**: Acidic fraction of EtOH extract of fresh, undried, winter leaves **Spectra**: UV, IR, ^{1}H NMR, MS **Derivative**: Acetate, m.p. 105°C, UV, IR, ^{1}H NMR, MS	*272*
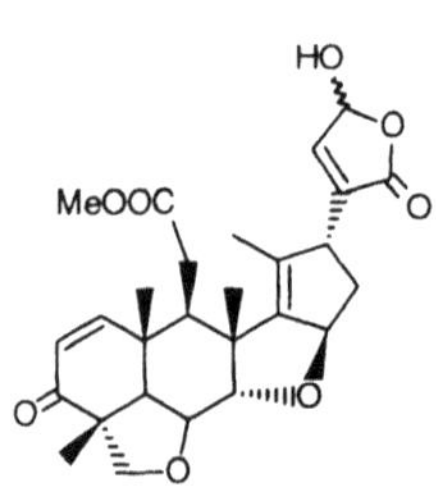 Margosinolide (**114**)	$C_{27}H_{32}O_8$; 484.209 **m.p.** 130°C (Plates, $CHCl_3$) $[\alpha]_D$ 50° ($CHCl_3$, c 0.02) **Isolation**: Acidic fraction of CH_2Cl_2 fraction of fresh, undried, uncrushed, spring twigs **Yield**: 25 mg/6 kg twigs (0.0007% on dry wt. Basis) **Spectra**: UV, IR, ^{1}H NMR, ^{13}C NMR, NOESY, MS **Derivative**: Acetate, m.p. 105–108°C, UV IR, ^{1}H NMR, ^{13}C NMR, MS	*255*

Table 13 *(continued)*

Salannin group		Ref.
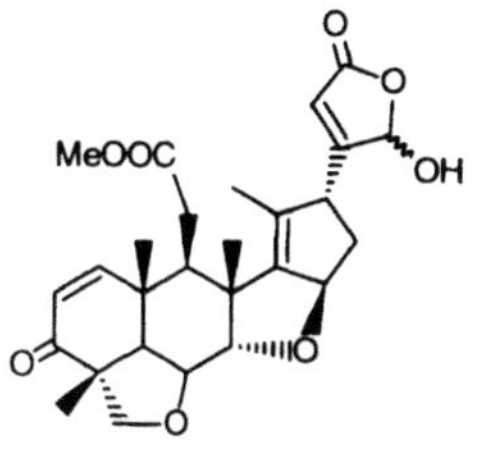 Isomargosinolide (**115**)	$C_{27}H_{32}O_8$; 484.2078 **m.p.** 125°C (rods, $CHCl_3$) $[\alpha]_D$ 14.28° ($CHCl_3$, c 0.07) **Isolation**: Acidic fraction of CH_2Cl_2 fraction of fresh, undried, uncrushed, spring twigs **Yield**: 50 mg/6 kg twigs (0.0014% on dry wt. Basis) **Spectra**: UV, IR, ^1H NMR, ^{13}C NMR, NOESY, MS **Derivative**: Acetate, m.p. 100–102°C, UV, IR, MS	*255*
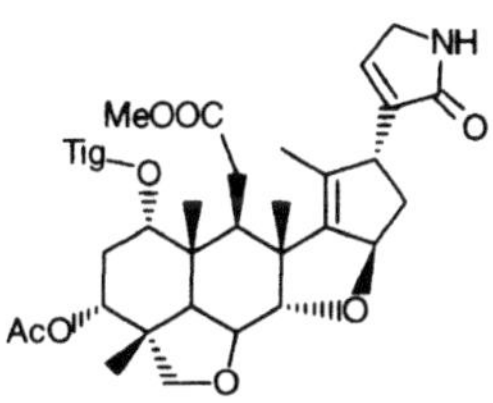 Salannolactame-21 (**116**)	$C_{34}H_{45}NO_9$; 611.3094 **m.p.** 213°C $[\alpha]^{\lambda}_{20^\circ C}\ \frac{589\quad 578\quad 546\quad 536\quad 405\quad 365}{121.8\ 126.8\ 147.2\ 269.7\ 334.5\ 462.7}$ ($CHCl_3$, c 0.1) **Isolation**: Petroleum ether – MeOH (1 : 1) extract of seed kernels **Yield**: 90 mg/3.6 kg seed kernels **Spectra**: IR, ^1H NMR, ^{13}C NMR, MS	*138*
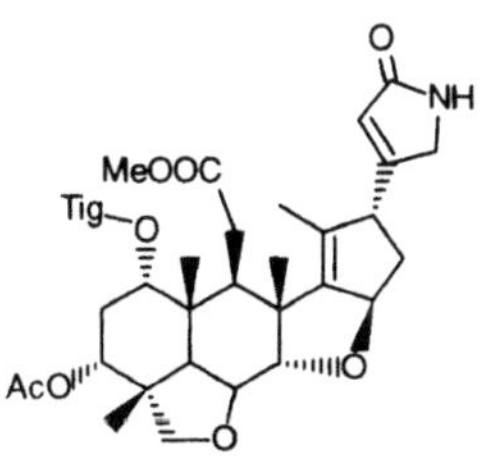 Salannolactame-23 (**117**)	$C_{34}H_{45}NO_9$; 611.3094 $[\alpha]^{\lambda}_{20^\circ C}\ \frac{589\quad 578\quad 546\quad 536\quad 405\quad 365}{126.3\ 133.8\ 153.2\ 285.3\ 353.8\ 495.0}$ ($CHCl_3$, c 0.1) **Isolation**: Petroleum ether – MeOH (1 : 1) extract of seed kernels **Yield**: 30 mg/3.6 kg seed kernels **Spectra**: UV, IR, ^1H NMR, MS	*138*

2.14. Azadirachtol Group

The structure of a substance named deacetylazadirachtinol from methanolic seed extracts (*139*) has been corrected to 3-tigloylazadirachtol (*36*, *114*, *128*) or azadirachtin B (**119**) (*222*, *223*) on the basis of detailed ^1H and ^{13}C NMR spectrometry. Some earlier assignments were corrected on the basis of homo-decoupling experiments and nuclear Overhauser effects in the Fourier transform difference spectrum.

2′,3′-Dihydrotigloylazadirachtol (**120**) and 3-isobutyrylazadirachtol (**121**) were isolated from methanolic extract of seeds (*131*). Azadirachtol (**118**), dihydro- (**122**) and tetrahydro-azadirachtols (**123**) have not been found in neem but were prepared chemically (*159*) and found to have comparable biological activity (*219*, *220*). 3-Tigloyl-13,14-desepoxy-17-hydroxy-azadirachtol or azadirachtin G (**124**) was reported as a minor component in neem seed extracts (*219*). 3α-Acetoxy-1α-hydroxyazadirachtol (**125**) has been isolated from the methanolic extract of seeds (*230*). Its structure was determined by comparison of its ^{1}H and ^{13}C NMR spectra with those of **119** and **135**. The ^{1}H NMR spectrum displayed signals of two olefinic protons at δ 5.06 and 6.44 for H-22 and H-23 of the dihydrofuran ring and a low-field singlet at δ 5.60 for H-21. Signals at δ 4.50, 1.70, 1.32 and 2.40 were characteristic of the four spin

Table 14

Azadirachtol group		Ref.
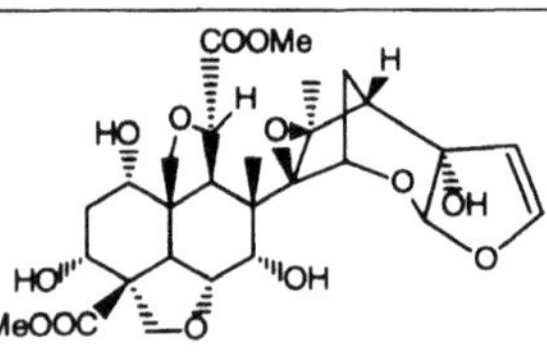 **Azadirachtol (118)**	Not isolated. Prepared by chemical modifications	*159, 218, 219*
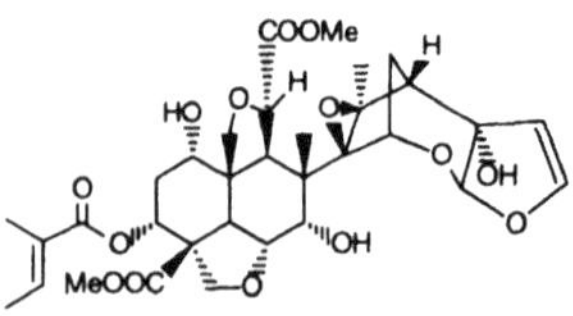 **3-Tigloylazadirachtol or Azadirachtin B (119)**	$C_{33}H_{42}O_{14}$; 662 **m.p.** 204–206°C (needles, EtOH-Water) $[\alpha]^{20}_{D}$ −69° (c 0.1) $[\alpha]^{\lambda}_{20}$ c: $\frac{589 \quad 578 \quad 546 \quad 486 \quad 405 \quad 365}{-69.4 \quad -70.8 \quad 77.7 \quad 125.0 \quad 151.4 \quad 204.2}$ (CH_2Cl_2, c 0.1) **Isolation**: Methanol extract of seeds **Spectra**: IR, UV, ^{1}H NMR, ^{13}C NMR, 2D Homo- (^{1}H-^{1}H) and Hetero- ^{1}H-^{13}C) COSY, MS **Activity**: same insecticidal activity as in azadirachtin and about 2.5 fold less active than azadirachtin as an insect growth inhibitor	*114, 128, 139, 140*
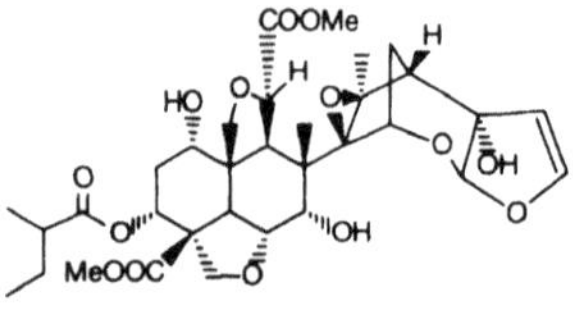 **2′,3′-Dihydrotigloylazadirachtol (120)**	$C_{33}H_{44}O_{14}$; 664.2731 **Isolation**: MeOH extract of seeds	*131*

Table 14 *(continued)*

Azadirachtol group		Ref.
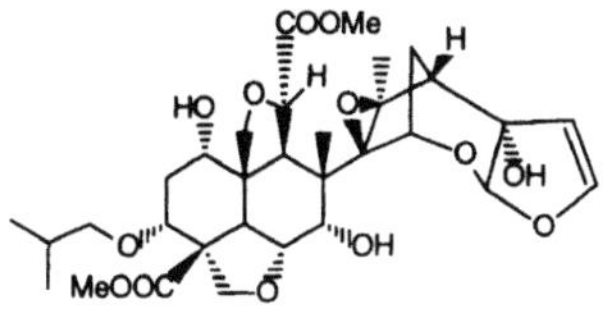 3-Isobutyroylazadirachtol (**121**)	$C_{32}H_{44}O_{13}$; 636.2781 **Isolation**: MeOH extract of seeds	*131*
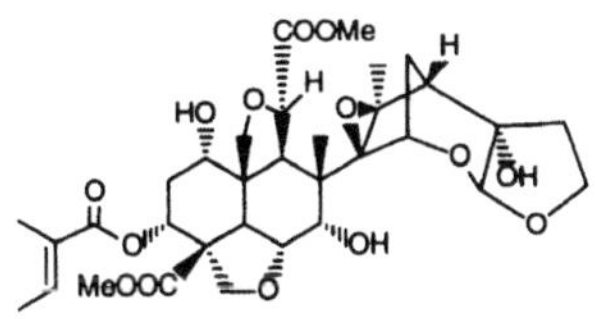 3-Tigloyl-22,23-dihydro-azadirachtol (**122**)	Not isolated from neem. Prepared by chemical modifications	*159, 218, 219*
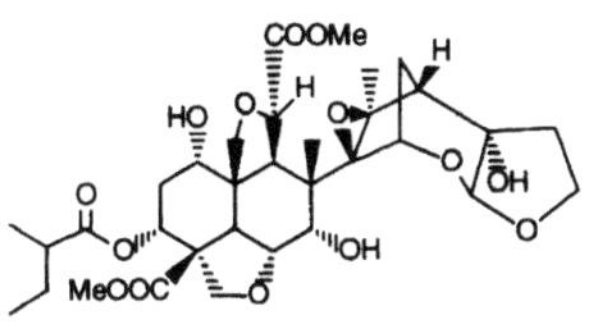 2',3'-Dihydrotigloyl-22,23-dihydroazadirachtol (**123**)	Not isolated from neem. Prepared by chemical modifications	*159, 218, 219*
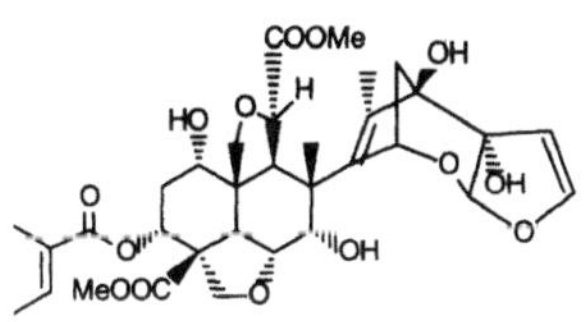 3-Tigloyl-13,14-desepoxy-17-hydroxyazdirachtol (Azadirachtin G) (**124**)	$C_{33}H_{42}O_{14}$; 662.2574 **Isolation**: Seed extract	*219*
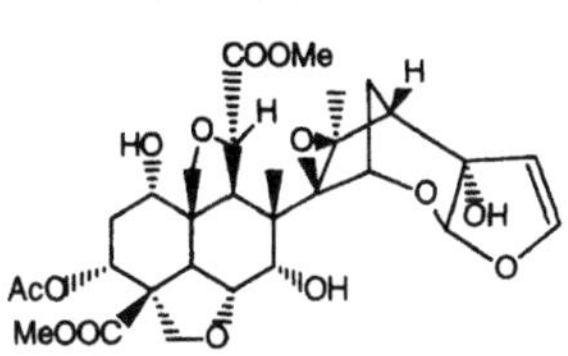 3α-Acetoxy-1α-hydroxy-azadirachtol (**125**)	$C_{31}H_{42}O_{14}$; 638.2574 **viscous liquid** **Isolation**: MeOH-Petroleum ether (1 : 1) extract of dried and powdered neem leaves **Yield**: 4 mg/10 kg seeds **Spectra**: ^{1}H NMR, ^{13}C NMR, MS	*230*

system H-15, Hα-16, Hβ-16 and H-17, while another four spin system with signals at δ 3.56, 2.38, 2.32 and 5.48 was that of H-1, H-2α, H-2β and H-3 and the AB system of 28-Hα,β had signals at δ 3.74 and 4.06. The ^{13}C signal of a hemiacetal carbon at δ 104.10 corresponding to C-11 of azadirachtin (**135**) was replaced in the spectrum of **125** by the tertiary carbon signal appearing at δ 79.62 was assigned to C-11.

2.15. Meliacarpin Group

The compounds of the meliacarpin group differ from the azadirachtols and azadirachtins by substitution at C-4 of a methyl group for the carbomethoxy group. 1-Tigloyl-3-acetyl-11-hydroxymeliacarpin or azadirachtin D (**127**) was isolated as gummy liquid from the methanolic extract of finely powdered seeds, the structure being assigned on the basis of ^{1}H NMR and ^{13}C NMR spectral data (*228*). Azadirachtin I (**128**) has been isolated from a HPLC peak of a azadirachtin-rich methanol extract of seed kernels. Structure **128** assigned to it on the basis of ^{1}H NMR, ^{13}C NMR and DEPT experiments and ^{1}H-^{1}H COSY spectral data, is very similar to that of **127** except for the lacks of a carbomethoxy group on C-11. Thus C-11 appeared as a doublet at δ 100.51. 11-H at δ 5.41 (d, $J = 4.4$ Hz) was coupled to 9-H at δ 3.21 (d, $J = 4.4$ Hz) (COSY). 3-H at δ 5.41 (dd, $J = 2.9$ and 2.4 Hz) was coupled to 2-Hα at δ 2.21 (ddd, $J = 8.0$, 3.4, 2.9 Hz) and to 2-Hβ at δ 2.26 (ddd, $J = 18.0$, 2.9, 2.4 Hz). 1,3-Diacetyl-11,19-deoxo-19-oxomeliacarpin (**130**), a possible intermediate in the biosynthesis of azadirachtin, was isolated as amorphous powder from the methanol extract of ground leaves (*137*), its structure being established by ^{1}H, ^{13}C NMR and NOE spectrometry. In contrast to the spectrum of azadirachtin (**135**) that of **130** exhibited four methyl signals instead of two and the signal of only one methoxycarbonyl group. The additional methyl groups were assigned to C-19 and C-29 by NOE experiments. Saturation of 19-H (δ 1.15) enhanced the signals of 30-H, 2-H_{β}, 6-H and 1-H, while enhancement of 19-H, 28-H_{β}, 6-H and 3-H was observed on irradiation of 29-H (δ 1.26) and 19-H, 6-H, 15-H and 7-H were enhanced on saturation of 30-H (δ 1.68).

2.16. Meliacarpinin and Azadirachtinin Group

The compounds of these groups are rearranged melicarpins and azadirachtins. 1-Tigloyl-3-acetyl-11-hydroxymeliacarpinin (**132**) was isolated from seeds (*131*) whereas 1-tigloyl-3-acetyl-11-methoxyazadir-

Table 15

Meliacarpin group		Ref.
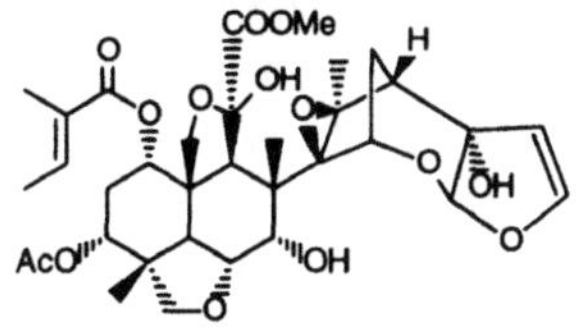 1-Tigloyl-3-acetyl-11-hydroxy-meliacarpin (Azadirachtin D) (**127**)	$C_{34}H_{44}O_{14}$; 676 **m.p.** gummy liquid **Isolation**: MeOH extract of finely powdered neem seeds **Yield**: 5 mg/10 kg neem seeds **Spectra**: ^{1}H NMR, ^{13}C NMR	*228*
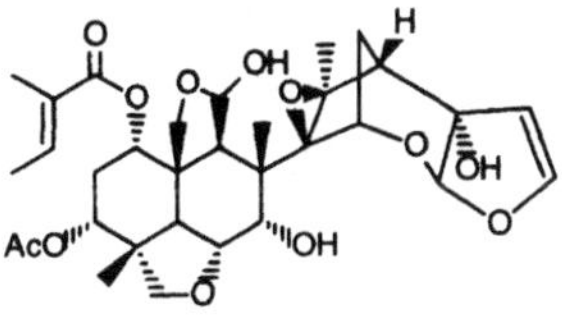 1-Tigloyl-3-acetyl-11-hydroxy-11-demethoxycarbonylmeliacarpin (Azadirachtin I) (**128**)	$C_{32}H_{42}O_{12}$; 618.2676 **m.p.** 198–200°C (EtOAc) $[\alpha]_D^{25}$ −21.8° ($CHCl_3$, c 0.8) **Isolation**: MeOH extract of neem seed kernels **Yield**: 3 mg/4 g extract **Spectra**: IR, UV, ^{1}H NMR, ^{13}C NMR, COSY, DEPT, FAB HRMS	*93, 94*
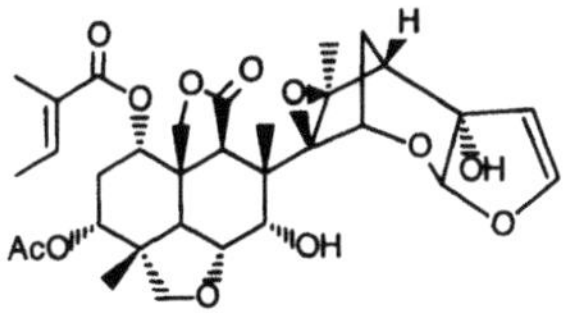 11-Demethoxycarbonyl-11-oxomeliacarpin (**129**)	$C_{32}H_{40}O_{12}$; 616.2519	*127*
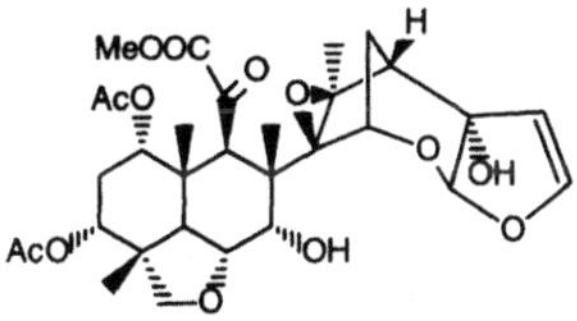 1,3-Diacetyl-11,19-deoxo-19-oxomeliacarpin (**130**)	$C_{31}H_{40}O_{13}$; 620 **m.p.** amorphous **Isolation**: MeOH extract of ground leaves **Yield**: 4 mg/5 kg neem seeds **Spectra**: ^{1}H NMR, ^{13}C NMR, NOE	*137*
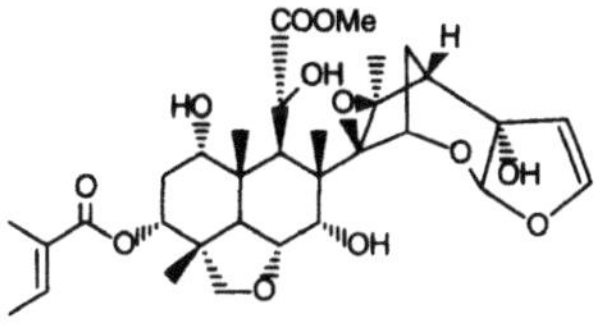 Azadirachtin F (**131**)	$C_{32}H_{44}O_{12}$; 620.2812	*219*

achtinin (**133**) was isolated from the methanolic extract of stem bark (*128*). The ^{1}H NMR spectrum of **133** showed signals of an acetoxy group, a tigloyloxy group, two methoxy substituents, and two OH groups determined by D_2O exchange. Two signals at δ 4.88 and δ 6.39 were assigned to be enol ether protons (22-H and 23-H respectively). On the basis of chemical shifts and multiplicities of the carbon signals two ester carbonyl groups, ten carbon atoms attached to oxygen by single bonds and two acetal carbons were identified in addition to the olefinic carbon atoms 22 and 23. 11-β-Hydroxyazadirachtinin (**134**) has been isolated from neem seeds (*231*), its structure being confirmed by ^{1}H NMR spectrometry which showed signals assignable to an acetoxy group, a tigloyl group and two carbomethoxy substituents. This compound showed close similarity to **132** and **133** with respect to the following signals (1) two signals to enol ether protons, (2) the four spin system 15-H, 16-H_α, 17-H. (3) signals for another four spin system 1-H, 2-$H_{\alpha,\beta}$ and 3-H, (4) An AB

Table 16

Meliacarpinin and azadirachtinin group		Ref.
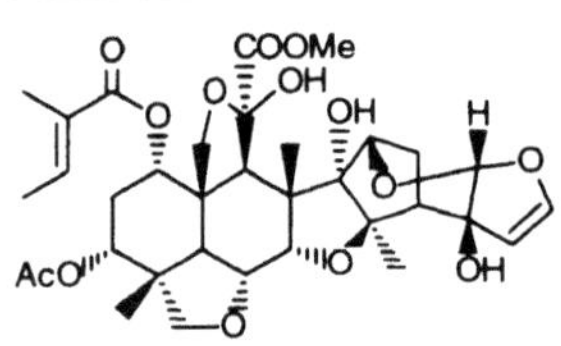 1-Tigloyl-3-acetyl-11-hydroxymeliacarpinin (**132**)	$C_{35}H_{46}O_{14}$; 690.2887 **Isolation**: seed extract	*131*
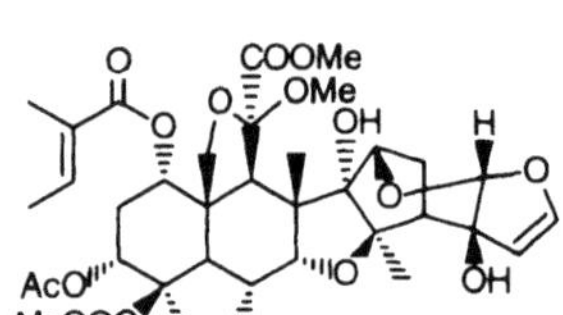 1-Tigloyl-3-acetyl-11-methoxyazadirachtinin (**133**)	$C_{36}H_{46}O_{16}$; 734 **m.p.** amorphous $[\alpha]^{\lambda}_{25°C}$ 589: −1.1, 578: −1.2, 546: −1.3, 436: −2.4, 404: −2.9, 365: −4.0, 313: −5.6 ($CHCl_3$, c 0.1) **Isolation**: Ether extract of neem bark **Yield**: 160 mg/20 kg bark **Spectra**: IR, ^{1}H NMR, ^{13}C NMR, EI-MS	*128*
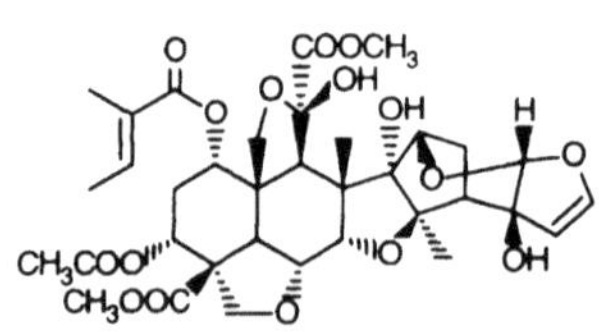 11β-Hydroxyazadirachtinin (**134**)	$C_{36}H_{46}O_{16}$; 734.278 **m.p.** 180–182°C (CH_2Cl_2-MeOH) **Isolation**: MeOH extract of seeds **Spectra**: ^{1}H NMR, ^{13}C NMR, FAB-MS	*141, 231*

system due to 28-$H_{\alpha,\beta}$, (5) the three spin system 5-H, 6-H and 7-H, and (6) two methyl signals assignable to 18-H and 30-H.

2.17. Azadirachtin Group

Azadirachtin (**135**) fascinated researchers working in the field of biologically active constitutents for over 25 years. Its biological activity, structure determination, reactions and synthesis are discussed in some detail at the end of this section. BUTTERWORTH and MORGAN (*44*, *72*) initiated the story of azadirachtin in 1968 by isolating it from neem seeds. Since then enormous amount of work has been carried out on its isolation, purification, structure determination, biological activity and biosynthesis (*1*, *2*, *92*, *128*, *139*, *151*, *160*, *176*, *244*, *245*, *304*, *306*, *313*, *315*, *316*, *317*). 3-Deacetylazadirachtin (**136**), 2′,3′-dihydrotigloylazadirachtin (**138**), 1-detigloyl-1-isobutyroylazadirachtin (**140**), 1-detigloyl-1-isovaleroylazadirachtin (**141**) and 1-detigloyl-1-isocaproyl-3-deacetyl-3-epoxymethacryoylazadirachtin (**142**) have also been extracted from seeds and identified by normal spectral procedures (*131*). 1-Detigloylazadirachtin (azadirachtin E) (**137**) is a hydrolysis product of azadirachtin (**135**) (*219*). Structure determinations of 3-deacetyl-3-cinnamoylazadirachtin (**139**) from powdered neem leaves (*128*) and of 22, 23-dihydro-23β-methoxyazadirachtin (Vepaol) (**143**) from neem seeds (*128*, *130*) have been carried with the help of ^{1}H NMR signals using homonuclear decoupling and measurements of the nuclear Overhauser effects (n.O.e) in the FT difference spectra. ^{13}C signals were assigned by means of two dimensional $^{1}H,^{13}C$ heteroscalar correlated spectra (*29*, *76*) H,C COLOC experiments (*111*) and the DEPT spectra (*30*). Isovepaol (**144**) is the 23α-isomer of **143**. Their ethoxy derivatives have also been prepared synthetically from azadirachtin and shown to be biologically active. 11-demethoxycarbonylazadirachtin (azadirachtin H) (**145**) has been isolated from neem seed kernels (*93*, *94*) by preparative HPLC. It is obvious from the NMR spectrum of **145** that it differs from **135** by having a hydrogen in place of a carbomethoxy group at C-11. H-11, appearing at δ 5.41 (d, $J = 4.4$ Hz), was found coupled to H-9 at δ 3.19 (d, $J = 4.4$ Hz), while the carbomethoxy group linked to C-11 in **135** and appearing at δ 3.68 (s), was absent in the spectrum of **145**. H-1 appeared at δ 5.36 (dd, $J = 2.9$, 2.4 Hz), a more normal position than in **135**, where at δ 4.75 (dd, $J = 2.9$, 3.1 Hz) it is apparently shielded due to conformational factors imposed by the presence of the carbomethoxy function on C-11. In the ^{13}C spectrum of **145**, C-11 appeared as a doublet at 99.57 ppm. The assignment of the C-3 acetate group was confirmed by a low power, selective frequency proton decoupling experiment. 1α-Destigloyl-1α-benzoylazadirachtin (**146**) was also isolated from neem seed, its structure being confirmed by ^{1}H NMR spectrometry (*231*). The proton signals in the spectra of **146** and **135** were strikingly similar except for the ester substituent at C-1. The characteristic signals attributable to the tiglate in **135** were missing and instead a mutiplet integrating for five protons at δ 7.32–7.88 attributable to aromatic protons was found, indicating the presence of a benzoyl group. As the chemical shifts of H-3 in the ^{1}H NMR spectra of both **135** and **146** were almost the same the benzoyl group could conveniently be placed at C-1 in **146**.

Azadirachtin K (**147**) has been isolated from the seed kernels and its structure established spectroscopically (*95*). The proton connectivities were established by ^{1}H-^{1}H COSY and decoupling experiments. ^{13}C NMR signals were assigned by DEPT experiments. Examination of the spectral data revealed that azadirachtins A (**135**) and K (**147**) were close analogues. In **135**, a five membered lactol system is formed by addition of the C-19 hydroxyl group to the C-11 carbonyl group whereas in **147**, lactone formation between the C-19 hydroxyl group and the C-11 carbomethoxy group (C-12) gives rise to a α-keto-δ-lactone system. 3-Deacetyl-11-desoxyazadirachtin (**148**) was isolated by reverse phase HPLC of methanolic neem seed extracts (*35*) which had been previously chromatographed on "floridin earth" to remove azadirachtin. The ^{1}H NMR spectrum proved to be very similar to that of deacetylazadirachtinol (**126**), indicating the structure was very closely related to **135**, the notable differences being the lack of acetate singlet at δ 1.95 coupled with the expected upfield shift of the 3-H proton to δ 3.52 and, more importantly, the increased complexity of the δ 4.0 to δ 4.5 region associated with structural change at C-11. Thus the hydroxyl singlet at δ 5.05 replaced by a broadened singlet at δ 4.47 which is coupled with the C-9 proton in the COSY 2-D spectrum. 11-Hydroxyazadirachtin B (**149**) has been isolated and identified from the seeds (*141*). Its ^{1}H and ^{13}C NMR spectra are similar to those of azadirachtin B (**119**) except for a few minor variations. Thus the signal of H-11 at δ 4.45 (d,<1 Hz) of **119** is absent, the signal of H-9 is shifted downfield (δ 3.34 compared with δ 3.17 in **119** and δ 3.34 in **135**) and the 30-methyl resonates at δ 1.73 (δ 1.45 in **119** and δ 1.74 in **135**). D_2O exchange and HCOH coupling studies in $(CD_3)_2SO$ indicated the presence of four hydroxyls (two secondary and two tertiary) *i.e.* one more hydroxyl than **119**.

11-Epi-azadirachtin H (**150**) has been isolated from the methanolic extracts of seeds (*211*). Its ^{1}H NMR spectrum exhibited the following signals: (i) Two olefinic protons H-22 and H-23 (δ 5.05 and 6.45) of the dihydrofuran ring and a low-field singlet for H-21 at δ 5.66; (ii) the four spin system H-15, H-16α,β and H-17 (δ 4.57, 1.68, 1.28 and 2.35); (iii) the four spin system H-1, H-2α, β and H-3 (δ 5.36, 2.39, 2.30 and 5.52); (iv) the AB system of H-28α, β and H-19α, β at δ 4.09 and 3.74; (v) the three spin system H-5, H-6 and H-7 (δ 3.36, 4.46 and 4.65); and (vi) the two methyl signals H-18 and H-30 (δ 1.98 and 1.32). The proton connectivities were established by ^1H-^1H COSY spectra and the multiplicities of all carbon signals were assigned by SEFT experiments. 13, 14-Desepoxyazadirachtin-A (**151**) has been isolated from the seed kernel extract by preparative HPLC extract, its structure being established by spectroscopic methods (*90*). The molecular formula of **151** differed from that of azadirachtin (**135**) only in having one less oxygen atom. The ^1H and ^{13}C spectra, differed significantly only in the chemical shifts of the methyl groups at C-13 (C-18) and C-8 (C-30) which appeared as singlets at 1.45 and 1.57 ppm compared with those of **135** at δ 2.10 and 1.74. When the epoxy oxygen bonding C-13-C-14 in azadirachtin (**135**) is absent both of these signals are shifted upfield. Both compounds have almost identical carbon resonances except for the signals at 92.3 ppm and at 94.5 ppm, due to the sp_2 carbons present in **151** and absent in **135**.

Table 17

Azadirachtin group		Ref.
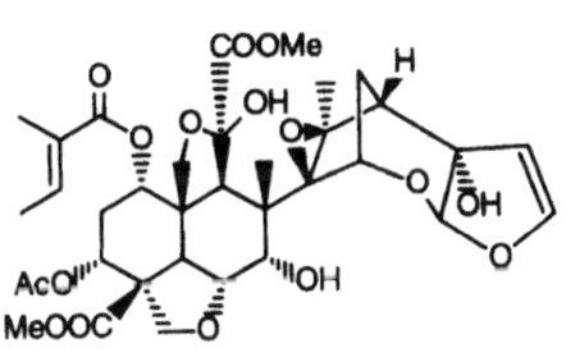 Azadirachtin (**135**)	$C_{35}H_{44}O_{16}$; 720	2, *91*, *128*, *139*, *160*, *176*, *245*, *306*, *315*, *316*, *317*
	m.p. 156°C; $[\alpha]_D$ −46°	
	$[\alpha]^{\lambda}_{20°C}$ 589 / −65.4; 578 / −69.8; 546 / −79.6; 436 / −135.2; 405 / −161.1 ($CHCl_3$, c 0.2)	
	Isolation: (i) Flash chromatography	*316*, *317*
	(ii) HPLC	*89*, *151*, *244*, *306*, *316*, *317*
	(iii) EtOH extract of fruits	*176*
	(iv) Super Critical Fluid Chromatography	*102*
	Derivatives: (i) 3-Deacetylazadirachtin, IR, ^1H NMR (ii) 11-O-Acetylazadirachtin, IR, ^1H NMR (iii) 11-O-Methylazadirachtin, IR, ^1H NMR (iv) 22,23-Dihydroazadirachtin, IR, ^1H NMR (v) 2′,3′,22,23-Tetrahydroazadirachtin, IR, ^1H NMR (vi) 1-Detigloyl-22, 23-dihydroazadirachtin, IR ^1H NMR (vii) 2′,3′-Dihydroxy-2′,3′,22,23-tetrahydroazadirachtin, IR, ^1H NMR (viii) 11,20-O,O-Dicarbomethoxy-22, 23-dihydroazadirachin, IR, ^1H NMR	

Table 17 *(continued)*

Azadirachtin group		Ref.
	Estimation in formulations	*314*
	Biosynthesis	*1, 132*
	Tissue culture	*2*
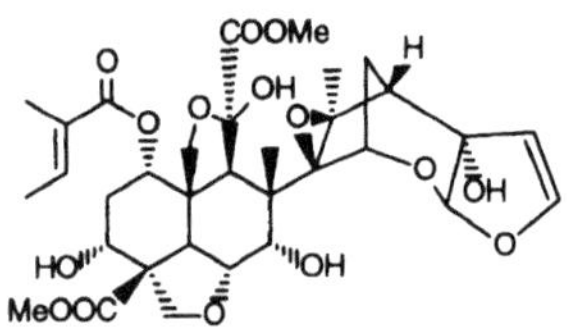 3-Deacetylazadirachtin (**136**)	$C_{33}H_{42}O_{15}$; 678.2523 **Isolation**: MeOH extract of seeds	*131*
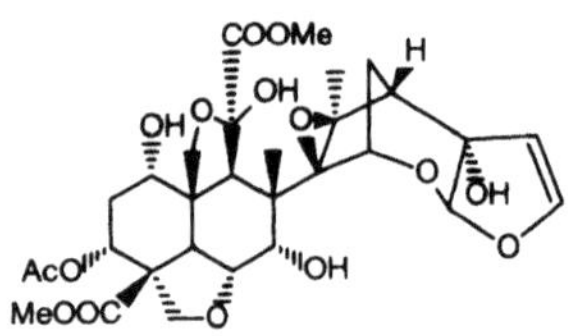 1-Detigloylazadirachtin (Azadirachtin E) (**137**)	$C_{30}H_{38}O_{15}$; 638.221 **m.p.** 638 **hydrolysis product of aza**	*219*
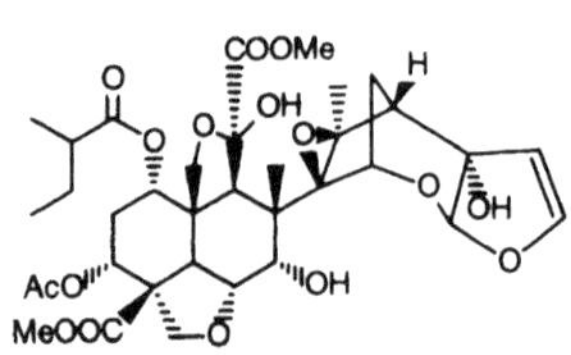 2',3'-Dihydrotigloyl-azadirachtin (**138**)	$C_{35}H_{46}O_{16}$; 722.2785 **Isolation**: seed extract	*131*
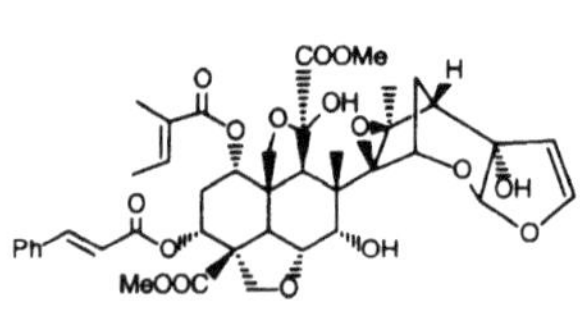 3-Deacetyl-3-cinnamoyl-azadirachtin (**139**)	$C_{42}H_{48}O_{16}$; 808 $[\alpha]^{\lambda}_{25°C}$ $\frac{589\ 578\ 546\ 436\ 405}{3.7\ 3.5\ 4.9\ 20.6\ 12.2}$ (CHCl$_3$, c 0.1) **Isolation**: Diethyl ether extract of defatted and powdered neem leaves **Yield**: 48 mg/326 g powdered leaves **Spectra**: UV, IR, ^{1}H NMR, ^{13}C NMR, MS	*128*

Table 17 *(continued)*

Azadirachtin group		Ref.
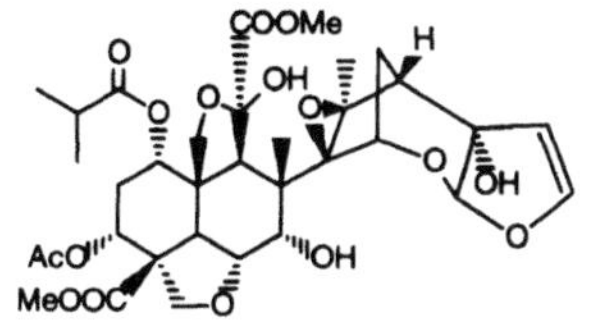 **1-Detigloyl-1-isobutyroyl-azadirachtin (140)**	$C_{34}H_{44}O_{16}$; 708.2629 **Isolation**: seed extract	*131*
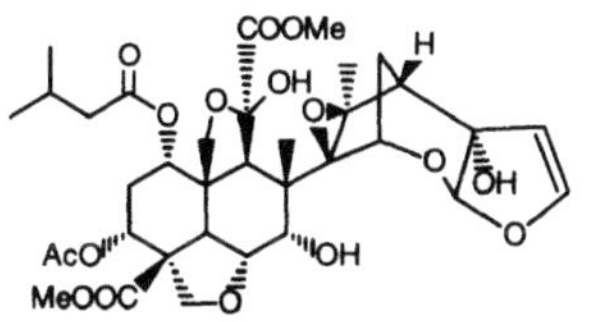 **1-Detigloyl-1-isovaleroyl-azadirachtin (141)**	$C_{35}H_{46}O_{16}$; 722.2785 **Isolation**: seed extract	*131*
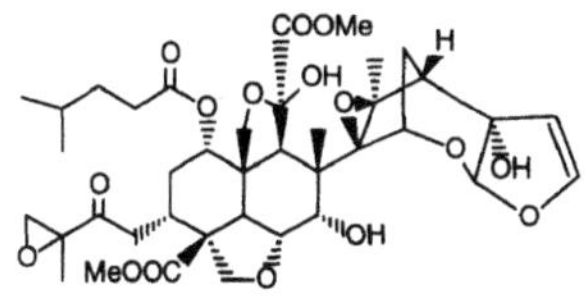 **1-Detigloyl-1-isocaproyl-3-deacetyl-3-epoxymethacroyl-azadirachtin (142)**	$C_{39}H_{52}O_{16}$; 776.3255 **Isolation**: seed extract	*131*
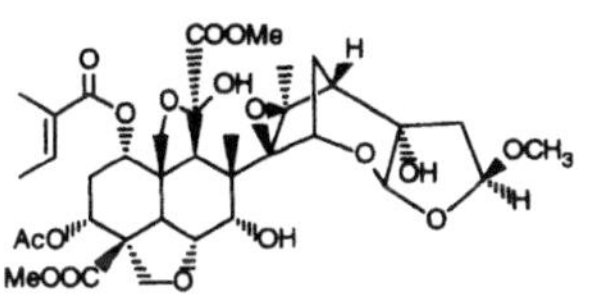 **22,23-Dihydro-23β-methoxy-azadirachtin (Vepaol) (143)**	$C_{36}H_{48}O_{17}$; 752.2891 **m.p.** amorphous $[\alpha]^{\lambda}_{20°C}$ $\frac{589 \quad 578 \quad 546 \quad 436 \quad 405 \quad 365}{-8.1 \quad -8.5 \quad -9.2 \quad -14.9 \quad -18.0 \quad -23.6}$ ($CHCl_3$, c 0.1) **Isolation**: MeOH extract of neem seed kernels **Yield**: 170 mg/3.6 kg neem seed kernels **Spectra**: IR, ^1H NMR, ^{13}C NMR, FDMS	*128, 130*
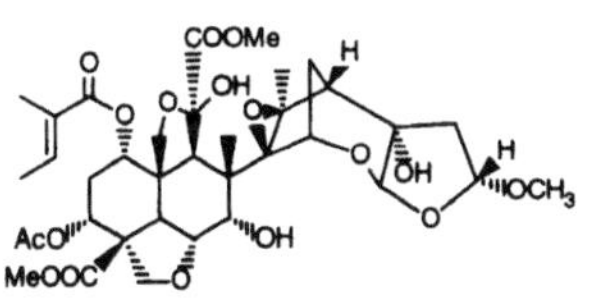 **Isovepaol (144)**	$C_{36}H_{48}O_{17}$; 752.2891	*233, 234*

Table 17 *(continued)*

Azadirachtin group		Ref.
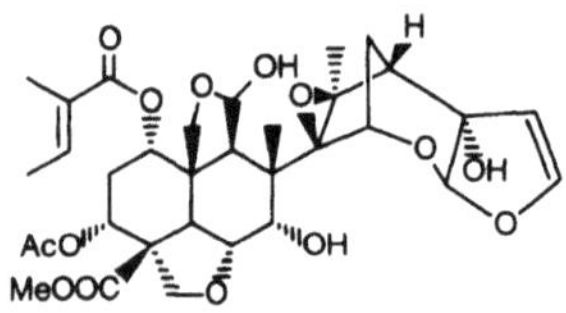 11-Demethoxycarbonyl-azadirachtin (Azadirachtin H) (**145**)	$C_{33}H_{42}O_{14}$; 662 **m.p.** 248° (HPLC peak dried at 75); 258–61°C (EtOAc) $[\alpha]_D^{25}$ −33.3° ($CHCl_3$, c 0.6) **Isolation**: MeOH extract of neem kernels **Yield**: 10 mg/4 g extract **Spectra**: IR, UV, ^{1}H NMR, ^{13}C NMR, COSY, DEPT, HR-FABMS	*93, 94*
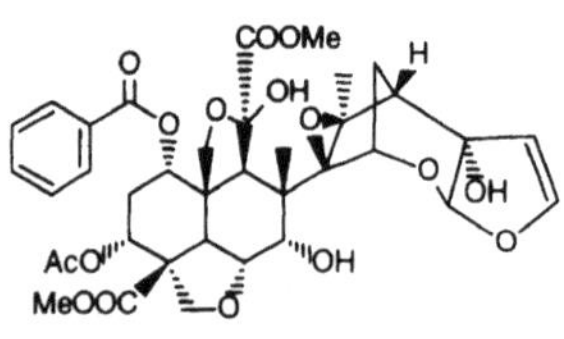 1α-Destigloyl-1α-benzoyl-azadirachtin (**146**)	$C_{37}H_{42}O_{16}$; 724 **Isolation**: MeOH extract of neem seeds **Spectra**: ^{1}H NMR	231
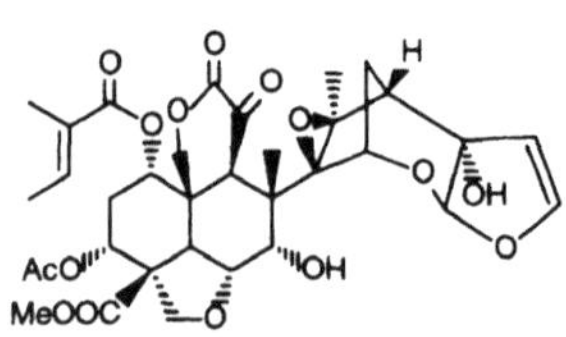 Azadirachtin K (**147**)	$C_{34}H_{40}O_{15}$; 688 **m.p.** 260°C **Isolation**: HPLC of extract of neem seed kernels **Yield**: 30 mg/4 kg seed kernels **Spectra**: IR, UV, ^{1}H NMR, ^{13}C NMR, COSY, DEPT, FAB-HRMS	*95*
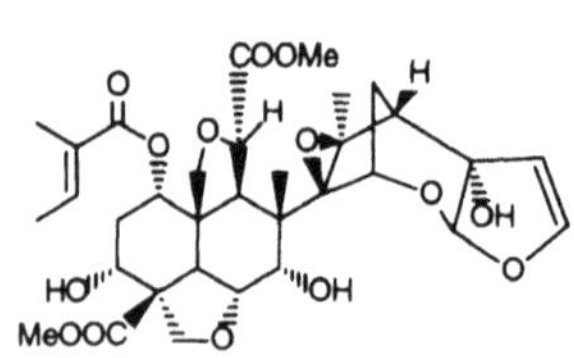 3-Deacetyl-11-desoxy-azadirachtin (**148**)	$C_{33}H_{47}O_{14}$; **m.p.** 149–51°C $[\alpha]_D^{20}$ −40.8° ($CHCl_3$, c 0.36) **Isolation**: EtOH extract of finely ground seeds **Spectra**: IR, ^{1}H NMR, ^{13}C NMR, MS	*35*
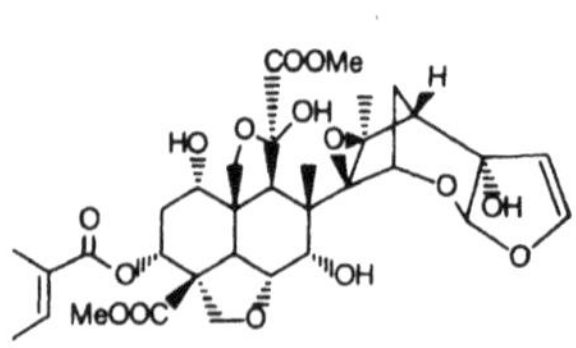 11-Hydroxyazadirachtin B (**149**)	$C_{33}H_{47}O_{15}$; **m.p.** 160–162°C **Isolation**: MeOH extract of dried and powdered neem seeds **Yield**: 10 mg/1 kg seeds **Spectra**: ^{1}H NMR, ^{13}C NMR, COSY, DEPT, negative FAB-MS	*141*

Table 17 *(continued)*

Azadirachtin group		Ref.
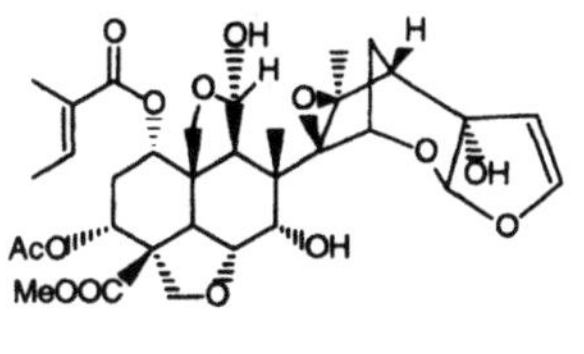 11-Epiazadirachtin H (**150**)	$C_{33}H_{42}O_{14}$; 662.2574 **m.p.** 170–73°C (microcrystalline solid) $[\alpha]_D^{20}$ 28.3° ($CHCl_3$, c 0.6) **Isolation**: MeOH extract of defatted neem seed kernels **Yield**: 30 mg/1.5 kg seeds kernels **Spectra**: IR, UV, ^{1}H NMR, ^{13}C NMR, ^{1}H-^{1}H COSY, NOESY, FABMS	*211*
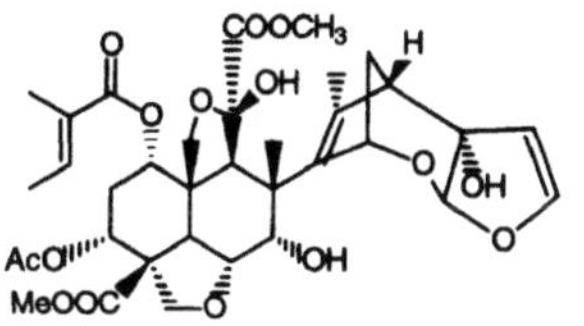 13,14-Desepoxyazadirachtin A (**151**)	$C_{35}H_{14}O_{15}$; 703.2619 **m.p.** 187°C $[\alpha]_D$ 13.64° ($CHCl_3$, c 0.44) **Isolation**: p-HPLC of seed kernel extract **Yield**: 23 mg / 500 g extract **Spectra**: IR, UV, ^{1}H NMR, ^{13}C NMR, FAB HRMS	*229*

2.18. Azadirachtin

2.18.1. Biological Activity

In traditional medicine products prepared from neem are used for treatment of diseases ranging from skin infections and cardiovascular disorders to diabetes and even cancer. Sodium nimbidinate was shown to be a potent diuretic (*33*) and produced a perceptible fall in blood pressure in anaesthetized rats (*80*). Nimbidin showed anti-arthritic properties (*201*), bestowed significant protection against gastric and duodenal lesions in animals and enhanced the healing process in acetic acid induced chronic gastric lesions in rats and dogs (*200*, *203*). Azadirachtin (**135**) is the most potent naturally occurring insect antifeedant against desert locusts at low concentration, is a growth disruptant and possesses anti-malarial activity. Its main advantage is that it is non-toxic for mammals, non-mutagenic and biodegradable. The antimalarial activity of crude neem extracts *in vitro* against *Plasmodium falciparum* was ascribed to gedunin (**45**) and nimbolide (**87**) (*305*). Later JONES and coworkers (*108*) found that azadirachtin (**135**), 22,23-dihydroazadirachtin (**156**), tetrahydroazadirachtin (**167**) and 22,23-dihydro-23-methoxyazadirachtin (**143**) inhibit formation of motile male gametes *in vitro*. Azadirachtin at a concentration of 100 μM, also completely inhibits exflagellation *in vitro* of the human malaria parasite, *P. falciparum*. However, the motility of fully formed male gametes is unaffected by azadirachtin. A comparison of active azadirachtin analogues with those found to be inactive suggested that the hemiacetal group at position C_{11} of azadirachtin is critical to antigametogenesist activity, many of the inactive synthetic analogues being substituted at this position. The decalin and hydroxydihydrofuran fragments of azadirachtin do not separately inhibit exflagellation at concentrations as high as 200 μM suggesting that either active sites on both fragments are required for activity or that the two fragments form an active site when linked. The fact that mixtures of the individual fragments are also inactive indicates that the spatial relationship within or between the active site(s) is critical to the compounds activity.

2.18.2. *Structure-Activity Relationships*

Systematic studies on structure-activity relationships in this area are rare and very little has been done so far in this field. Pharmacological tests have been carried out mainly with crude extracts and in relatively few cases with pure compounds (*280, 311*). It is difficult to discuss such relationships on the basis of the available data because different test organisms, different set-ups of bio-assays and different structure types have been used by the various investigators. Hydrogenation of the olefinic bond of the dihydrofuran ring of azadirachtin has been reported (*222*) to have little or no effect on its metamorphosis-inhibiting activity against *Epilachna varivestis* and *Locusta migratoria*. Azadirachtin and its deacetyl derivative, were fully active as antifeedants against the desert locust (*Schistocerca gregaria*) (*175*). Acetylation and/or trimethylsilylation of the hydroxyl groups of **135** gave less active antifeedants against *S.gregaria*. However, in another report (*315*) neither deacetylation nor hydrogenation of the two carbon-carbon double bonds of **135** had any significant effect on growth-inhibitory and lethal activities against the major agricultural insect pest *Heliothis virescens*. Removal of the tigloyl group resulted in a moderately less active derivative. However, converting the relatively hydrophobic tigloyl group to the hydrophilic α,β-dihydroxy-α-methylbutyryl moiety caused a dramatic reduction in activity. The greatest losses of activity were observed when the hydroxyl groups were modified, either by carbomethoxylation or by *O*-methylation. These results showed that the hydroxyl groups in azadirachtin are essential for activity and that, for maximum activity, the molecule must also have a lipophilic region.

Another report suggested that the hydroxyfuranacetal moiety is clearly important for high levels of potency as compounds not possessing this structural feature, like salannin (**107**), do not affect insect development to the same extent. On the other hand, the ester group attached to C-4 seems to be essential for the antifeedant activity of azadirachtin, since all meliacarpin derivatives (**127–131**) isolated so far, in which this ester group is replaced by a methyl group, have been found to be insect growth regulators but exhibit no or very low antifeedant activity (*125, 126, 131*). The stereochemistry at C-7 is crucial and the bridging oxygen substituent at C-6 may play some role. The functionalities at C-11 position might have an important role to play in azadirachtin's biological activities (*159*). Still this field needs further studies.

2.18.3. Structure Determination

Azadirachtin was isolated for the first time by MORGAN and BUTTERWORTH in 1968 (*44*) who suggested two partial structures (**152**) and (**153**) on the basis of functional group analysis by NMR studies (*46*) and determined its molecular formula as $C_{35}H_{44}O_{16}$ (*45*). On the basis of selenium dehydrogenation of the product obtained by $LiAlH_4$ reduction, which yielded di-, tri- and possibly tetramethylnaphthalenes but no phenanthrenes, the authors concluded that azadirachtin belonged to a new class of C-*seco*-hexanortriterpenoid related biogenetically to the C-*seco*-tetranortriterpenoids, salannin (**107**) and nimbin (**76**). Later on, in 1975, NAKANISHI and co-workers (*181, 319*) proposed the first complete structure (**154**) of azadirachtin based on partially relaxed Fourier transform (PRFT) ^{13}C NMR spectroscopy, continuous wave decoupling (CWD) and a hypothetical relationship of azadirachtin with salannin (**107**) and nimbin (**76**). The authors established the presence of one acetate, one tiglate, two methoxycarbonyls, two olefinic protons in addition to that in the tiglate and two quarternary methyls. The tiglate and acetate residues were assigned to C-1 and C-3, respectively, while the angular C-19 methyl group was appended to C-10.

After a long silence lasting until 1984, KUBO et al. (*139, 140*) reported the isolation and structure determination of deacetylazadirachtinol (**155**), a new congener of azadirachtin, for which they proposed formula **155** with a single C(8)-C(14) bond joining two halves. This conclusion was a major breakthrough. For the first time it was shown that azadirachtin might contain two distinct parts – the hydroxyfuran fragment and the decalin portion both joined by C(8)-C(14) bond. Soon

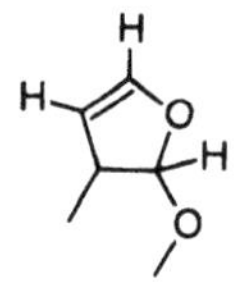

152

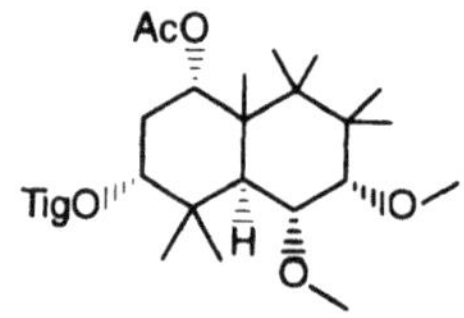

153

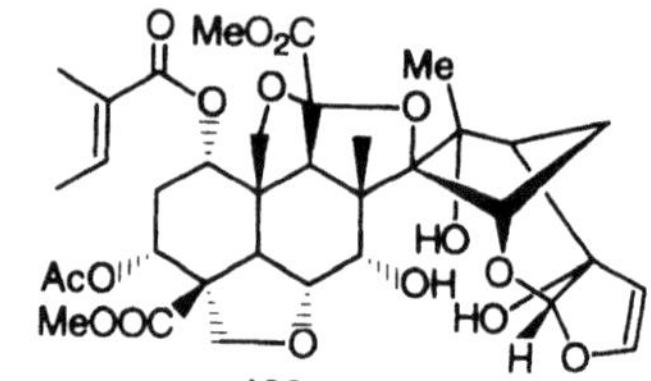

126

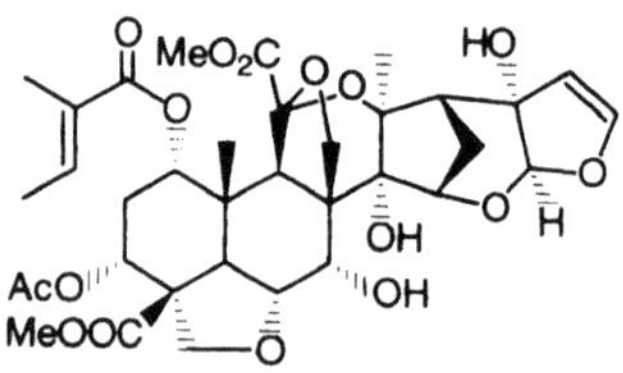

154

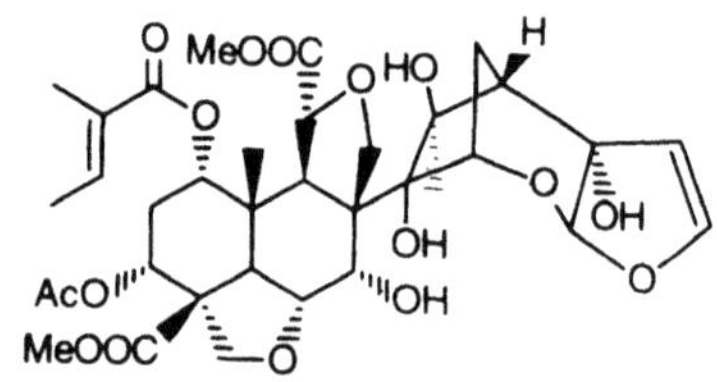

155

after, in early 1985, the structure of azadirachtin was revised to **126** by Ley and co-workers (*34*) using 2D-NOESY and 1D-NOE difference spectroscopy, but more or less simultaneously Kraus *et al.* (*122*, *123*, *129*, *130*) proposed structure **135** which is correct. The correct tetrahydrofuran annulation locus (*81*) and the presence of the C-11 hemiacetal (*206*) was confirmed by using one-dimentional NOE-difference spectroscopy in conjunction with ^{13}C deuterium isotope experiments. A single crystal X-ray analysis by the Ley group (*40*) in early 1986 verified structure **155**. The story of the azadirachtin structure has been used as an example of the methodology of structure determination by Taylor (*297*). Ley and co-workers (*163*) established the absolute configuration of the left hand fragment of azadirachtin by the modified Mosher method and used azadirachtin as an example for the use of the Structure Determination (Lsd) program assisted NMR analysis for determination of structure.

2.18.4. Chemical Reactions

Azadirachtin is acid, base, and light sensitive due to the plethora of oxygen functions which include an acetal, a hemiacetal, an enol ether, a tetra-substituted oxirane, secondary and tertiary hydroxyl groups, the ether function of a tetrahydrofuran as well as various carboxylic esters. The early work on the structure of this complex molecule depended mainly on chemical reactions involving these functions. These are listed in Table 18.

References, pp. 132–149

Table 18

Compound	Reagents and conditions	Product
A. Acetylation		
1. Azadirachtin (**135**)	Acetic anhydride, Pyridine	11-Acetoxyazadirachtin (**157**)
2. 22,23-Dihydroazadirachtin (**156**)	Ac_2O, Et_3N, DMAP, CH_2Cl_2, r.t.	11-Acetoxy-22,23-dihydroazadirachtin & 11,20-Diacetoxyazadirachtin (**158**)
3. Azadirachtin (**135**)	$(MeO)_2CO$, 75°C, 30 min	11,20-Dicarbomethoxyazadirachtin (**159**)
4. 3-Tigloylazadirachtol (**119**)	Ac_2O, Et_3N, DMAP, CH_2Cl_2	20-Acetyl-3-tigloylazadirachtol (**160**)
B. Silylation		
1. Azadirachtin (**135**)	*bis*-Trimethylsilylacetamide (BSA), 10 min.	11,20-*bis*-Trimethylsilylazadirachtin (**161**)
	BSA, Trimethylsilylimidazole, TMS Cl, Py, 60°C, 90 h	7,11,20-*tris*-Trimethylsilylazadirachtin (**162**)
2. 3-Tigloylazadirachtol (**119**)	TMSOTf, Et_3N, CH_2Cl_2	1,7,20-*tris*-Trimethylsilylazadirachtin (**163**)
C. Methylation		
1. Azadirachtin (**135**)	Ag_2O, MeI, Δ, 3 h	11-Methoxyazadirachtin (**164**)
	above reagents on prolonged treatment	11,20-Dimethoxyazadirachtin (**165**)
2. Azadirachtin (**135**)	KOH, MeI, 5 min	7,11,20-Trimethoxyazadirachtin (**166**)
D. Hydrogenation		
1. Azadirachtin (**135**)	Pd/C, H_2, MeOH, 2 h	22,23-Dihydroazadirachtin (**156**)
	Pd/C, H_2, MeOH, 6 h	2′,3′,22,23-Tetrahydroazadirachtin (**167**)
2. 3-Tigloylazadirachtol (**119**)	Pd/C, H_2, MeOH, 10 min	2′,3′,22,23-Tetrahydroazadirachtol (**168**)
E. Reactions of Enol Ether function		
1. Azadirachtin (**135**) or 11-Methoxyazadirachtin (**164**) or	AcOH, r.t., 72 h	23-anomeric acetates (**169**) ($\alpha:\beta$ 2 : 1)

Table 18 *(continued)*

Compound	Reagents and conditions	Product
Azadirachtol (**118**) or 3-Tigloylazadirachtol (**119**)		
2. Azadirachtin (**135**)	Br_2, MeOH, O°C, 5 min	*trans*-Bromo acetals at C-22,23 (**170**)
3. Azadirachtin (**135**)	Lewis acids	Azadirachtinins
F. Saponification Reactions		
1. 3-Tigloylazadirachtol (**119**)	Et_3N, MeOH, H_2O (1 : 5 : 1), 65°C	Azadirachtol (**118**)
2. Azadirachtin (**135**)	MeOH, KOH	3-Desacetylazadirachtin (**136**)
3. Azadiracthin (**135**)	Sod. Periodate, $KMnO_4$	Detigloylazadirachtin (**137**)
4. 22,23-Dihydroazadirachtin (**156**)	(i) a. O_3, MeOH, −50°C b. Aq. $NaHCO_3$ (ii) OsO_4, aq. MeCN, 25°C, 30 min then $NaIO_4$, 230 min	1-Pyruvic-22,23-dihydroazadirachtin (**171**) 1-Detigloyl-22,23-dihydroazadirachtin (**172**) *cis-vicinal*-diol (**173**)
G. Azadirachtol Chemistry		
1. Azadirachtol (**118**) or 22,23-Dihydroazadirachtol or 23-α/β-Acetoxy-22,23-dihydroazadirachtol (**174**)	2-Methoxypropene, PPTS, CH_2Cl_2, r.t., 3 h	1,3-Acetonides (**175**)
2. Azadirachtol (**118**)	Ac_2O, Et_3N, DMAP, CH_2Cl_2, 2 d, r.t. Acetic pivalic anhydride, Et_3N, CH_2Cl_2, 0°C, r.t., 13 h	1,3,20-Triacetoxyazadirachtol (**176**) 3-Acetoxyazadirachtol (61%) (**177**) & 3,20-Diacetoxyazadirachtol (6%) (**178**)

H. Oxidation Reactions		
1. 23α-Acetoxy-22,23-dihydro-11-methoxy-azadirachtin (**179**)	Pyridinium dichromate (PDC), 4 A° sieve, CH_2Cl_2, r.t., 40 h	23α-Acetoxy-22,23-dihydro-7-keto-11-methoxyazadirachtin (**180**)
	175°C, 1.7×10^{-3} mm Hg, 5 min	7-Keto-11-methoxyazadirachtin (**181**)
2. Azadirachtin (**135**)	Periodinane, CH_2Cl_2, r.t., 30 min	23-Acetoxy-$\Delta^{20,22}$-azadirachtin (**182**)
I. 7-Ketoazadirachtins		
1. 7-Keto-11-methoxyazadirachtin (**181**)	Et_3N, MeOH, H_2O, (1 : 5 : 1), 65°C, 32 h	3-Desacetyl-1-detigloyl-7-keto-11-methoxyazadirachtin (**183**)
J. Retro-aldol Reaction		
1. 22,23-Dihydro-11-methoxyazadirachtin (**184**)	PDC, CH_2Cl_2	compounds **185** and **186**
Compound **186**	NaOMe, MeOH, r.t., 24 h	degradation products (**187** & **188**)
2. 11-Acetoxy-22,23-dihydro-azadirachtin (**192**)	PCC, 4 A° sieves, CH_2Cl_2, 48 h, r.t. then 35°C, 24 h	Ketocarbonate **194** (49%) and keto-compound **193** (11%)
Ketocarbonate (**194**)	5% Et_3N, MeOH, Δ, 30 h then CH_2Cl_2 or Et_3N, MeOH, H_2O (1 : 5 : 1), r.t., 24 h then CH_2N_2, CH_2Cl_2	Decalins (**195**, **196** and **197**)

2.18.4.1. Reactions of –OH Group

2.18.4.1.1. Acetylation

Azadirachtin on treatment with acetic anhydride afforded a tertiary acetate (**157**) (*46*, *128*, *315*, *319*) which was later identified as a C-11 derivative by KRAUS *et al.* (*128*). The formation of the 11-acetate along with the C-11,C-20 diacetate (**158**) from 22,23-dihydroazadirachtin (**156**) (*161*), and 11,20-dicarbomethoxyazadirachtin (**159**) from azadirachtin (**135**) (*315*) reveals the order of reactivity C(11)OH > C(20)OH ≫ C(7) OH for azadirachtin.

3-Tigloylazadirachtol (**119**) serves as a substrate for comparing the reactivities of the axial C(1)OH and the *endo*-C(20)OH groups. The acetylation is highly regioselective in the presence of the tiglate residue giving 20-acetyl-3-tigloylazadirachtol (**160**) exclusively (*154*).

2.18.4.1.2. Silylation

Both *bis*- and *tris*-trimethylsilylethers (**161**, **162**) were prepared by MORGAN and BUTTERWORTH (*45*, *46*) for mass spectrometry; however, silylation of the secondary C(7)OH group required harsh conditions which proved the unreactivity of the -OH group at C-7. Acetylation, silylation and oxidation of the tertiary hydroxyls (at C-11 and C-20) in preference to the secondary hydroxyl at C-7 indicated that the latter is in a highly hindered position. Similarly, 3-tigloylazadirachtol (**119**) afforded 1,7,20-*tris*-trimethylsilyazadirachtol (**163**) (*156*).

2.18.4.1.3. Methylation

Methylation of azadirachtin (**135**) with iodomethane under Purdie conditions was earlier reported to be low yielding and difficult (*315*); however, a slight modification of the procedure proceeded efficiently (*162*) to form 11-methyoxyazadirachtin (**164**). Prolonged treatment afforded 11,20-dimethoxyazadirachtin (**165**) (*158*) comparable to acylation and silylation. Formation of 7,11,20-trimethoxyazadirachtin (**166**) had been reported by NAKANISHI and co-workers (*304*).

2.18.4.2. Hydrogenation

The instability of azadirachtin may be attributed in part to the presence of the tiglate and enol ether functions. However, azadirachtin was resistant to hydrogenation under a variety of conditions. Reduction

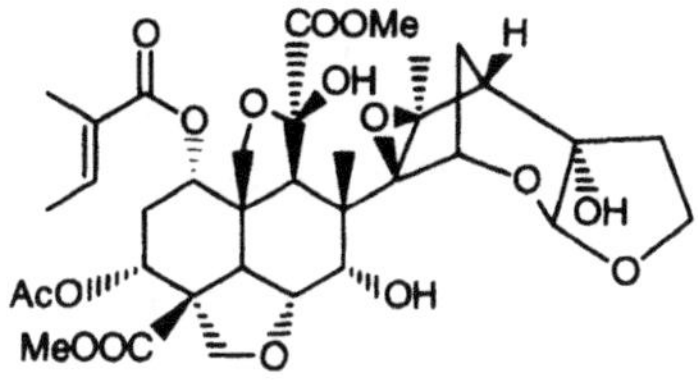

22, 23-Dihydroazadirachtin (**156**)

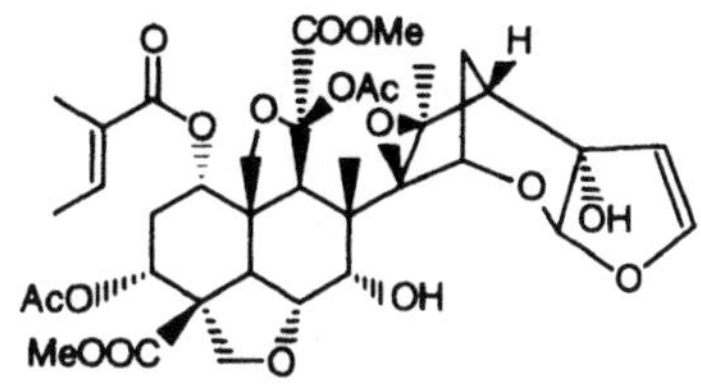

11-Acetoxyazadirachtin (**157**)

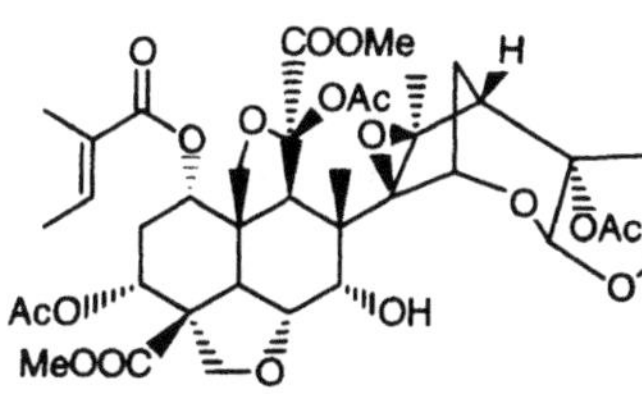

11,20-Diacetoxy-22,23-dihydroazadirachtin (**158**)

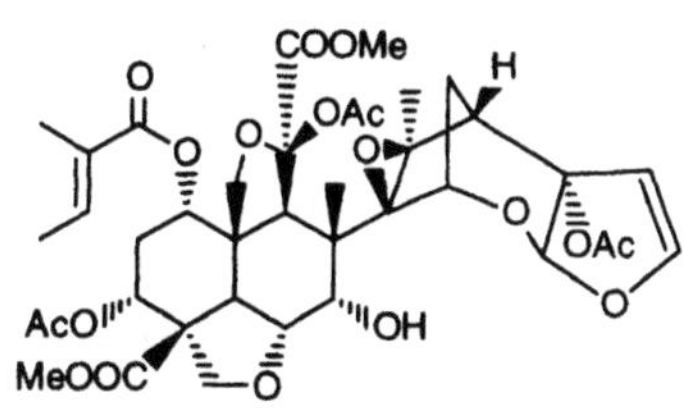

11,20-Dicarbomethoxyazadirachtin (**159**)

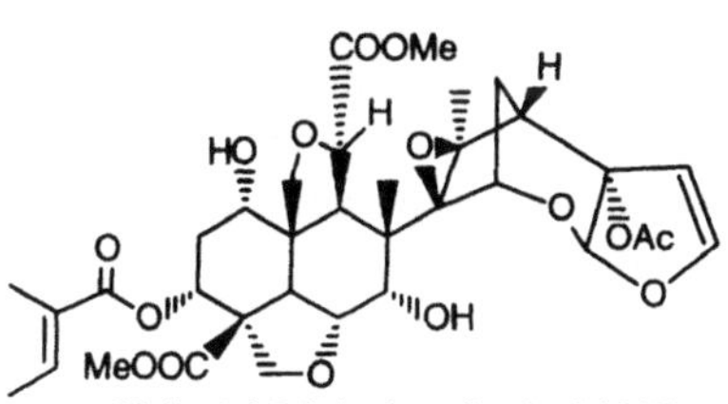

20-Acetyl-3-tigloylazadirachtol (**160**)

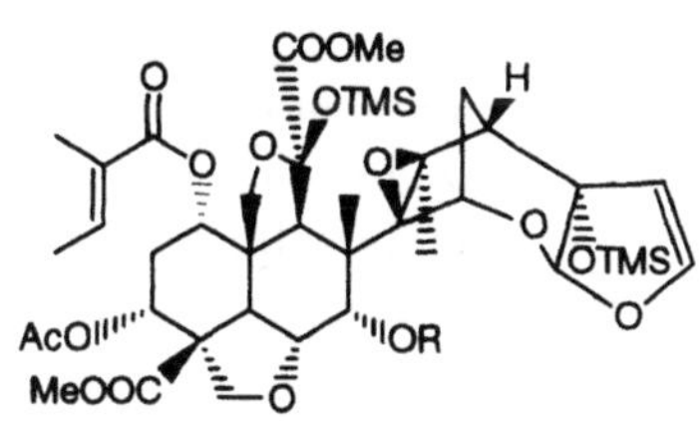

R = H 11,20-*bis*-Trimethylsilylazadirachtin (**161**)
R = TMS 7,11,20-*tris*-Trimethylsilylazadirachtin (**162**)

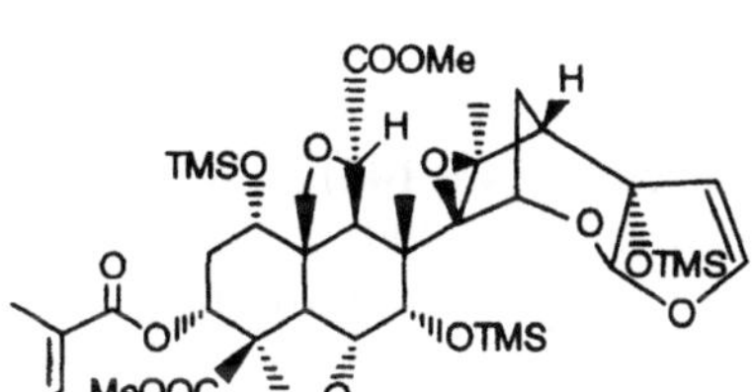

1,7,20-*tris*-Trimethylsilylazadirachtol (**163**)

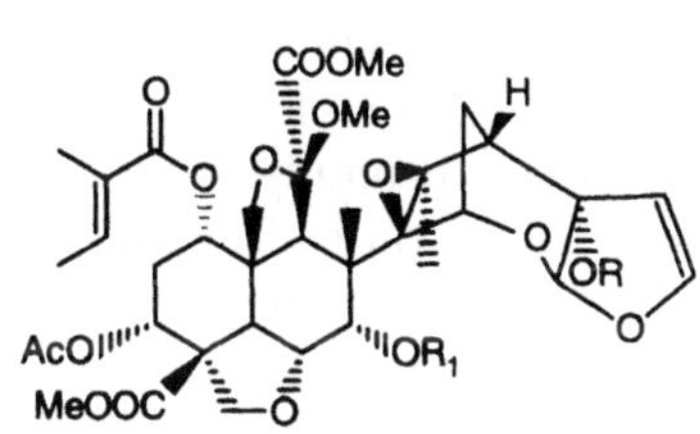

R=H, R_1=H 11-Methoxyazadirachtin (**164**)
R=Me, R_1=H 11,20-Dimethoxyazadirachtin (**165**)
R=R_1=Me 7,11,20-Trimethoxyazadirachtin (**166**)

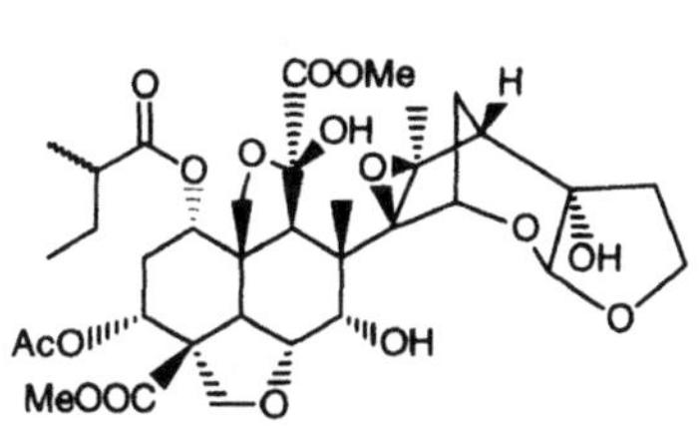

2',3',22,23-Tetrahydroazadirachtin (**167**)

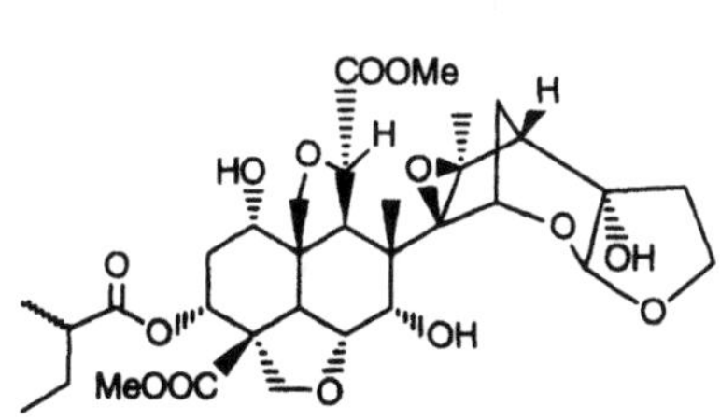

2',3',22,23-Tetrahydro-3-tigloylazadirachtol (**168**)

of the enol ether function (*46*) was successfully accomplished with Adam's catalyst (Pt_2O) at 50 lb/in^2 without any loss in biological activity (*217*). Under the same conditions the tiglate moiety was reduced as well (*315*). LEY and co-workers found that azadirachtin is selectively reduced using palladium on charcoal to either 22,23-dihydroazadirachtin (**156**) (*36*) or 2′,3′,22,23-tetrahydroazadirachtin (**167**) (*156*). 11-Methoxyazadirachtin (**164**) may also be hydrogenated under same conditions, yielding a substance with greatly improved stability and without loss of insect antifeedant activity (*158*).

Hydrogenation of 3-tigloylazadirachtol (**119**) giving 2′,3′,22,23-tetrahydro-3-tigloylazadirachtol (**168**) is, interestingly, neither chemo- nor stereo-selective (*36*, *156*), perhaps reflecting a less constrained environment for the C-3 tiglate residue in this molecule.

2.18.4.3. Reactions of the Enol Ether Functions

The acid catalysed rearrangement of azadirachtin by acetic acid afforded a mixture of anomeric acetates (**169**) (*156*, *157*) which could be reconverted to azadirachtin by pyrolytic *syn*-elimination of acetic acid. However, other carboxylic acids like formic and propionic acid did not produce the desired derivatives. The enol ether functions of related compounds including 3-tigloylazadirachtol (**119**), 11-methoxyazadirachtin (**164**) and azadirachtol (**118**) were protected in a similar way.

Treatment of azadirachtin (**135**) with a various Brønsted acids in the presence of methanol led complex mixtures of azadirachtinins, while treatment of azadirachtin with a methanolic solution of bromine afforded an isolable mixture of *trans*-bromoacetals (**170**) (*157*), which were converted to the naturally occurring methyl acetals (**143**, **144**) (*128*, *130*, *233*) with tributyltin hydride in refluxing benzene (*50*, *300*).

The successful formation of 1-tigloyl-3-acetyl-11-methoxyazadirachtinin (**133**) is encouraging and implies that choice of a suitable Lewis acid (*240*) to avoid rearrangement will allow this protocol to be extended to azadirachtin synthesis. The loss of stereochemical integrity at C-11 upon rearrangement imply a specific role for the C(13)-C(14) oxirane in fixing the relative configuration of this stereogenic centre through a hydrogen bond.

2.18.4.4. Saponification Reactions

Azadirachtin on saponification with methanol under basic conditions (*46*, *315*) led to the formation of 3-deacetylazadirachtin (**136**) in low yield. Removal of the intrinsically less reactive and sterically encumbered C-1 tiglate residue is difficult. The C-1 detigloyl compound (**172**) can be obtained by preparing a highly reactive C-1 pyruvate (**171**) followed by hydrolysis. The chemoselective oxidation of the tiglate in presence of the enol ether of azadirachtin is accomplished using sodium periodate and potassium permanganate (*46*), while in the case of 22,23-dihydroazadirachtin (**156**), sodium periodate (*36*), ozone (*154*), and osmium tetroxide (*315*) have all been used. The C-2′, C-3′ *cis-vicinal*-diol (**173**) may also be isolated in the later case (*315*).

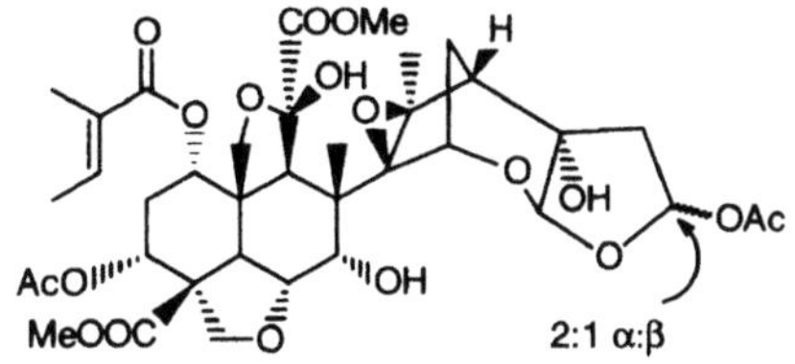

23α/β-Acetoxy-22,23-dihydroazadirachtin (**169**)

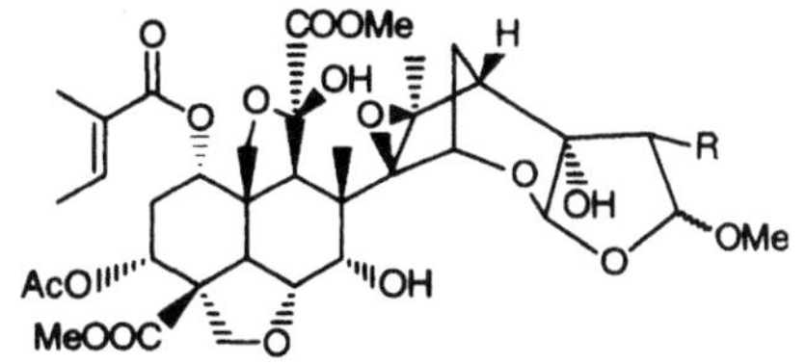

170 R=Br
143 (β) : **144** (α) 3 : 1 R=H

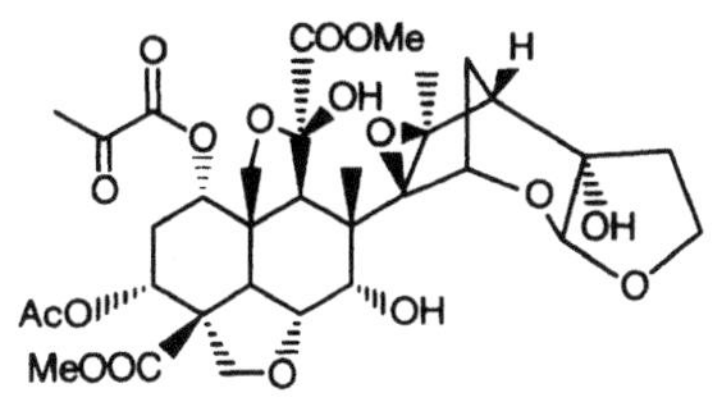

171

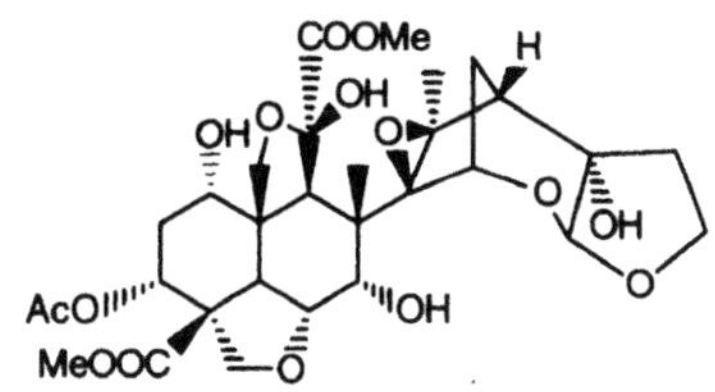

172

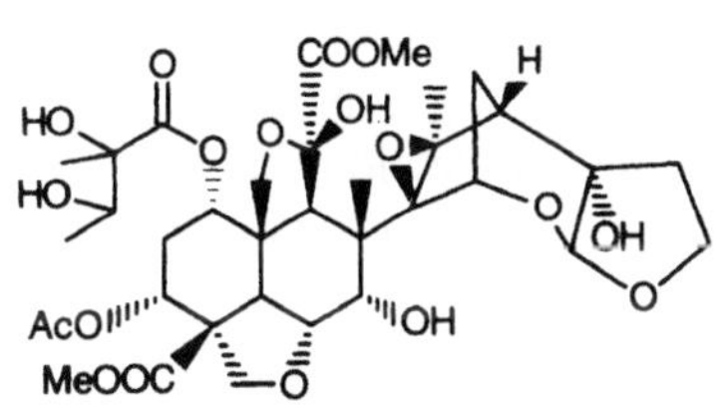

173

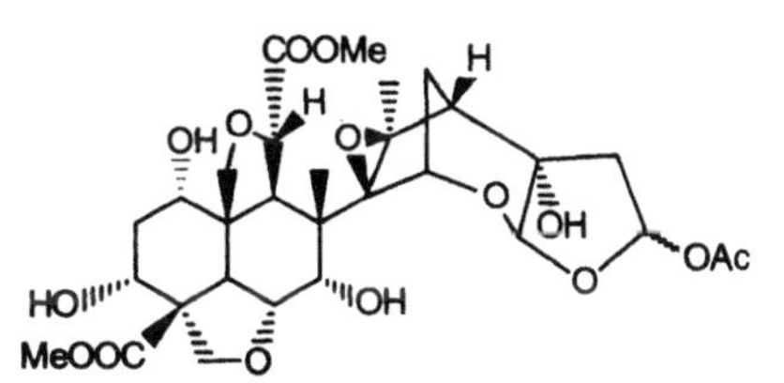

174

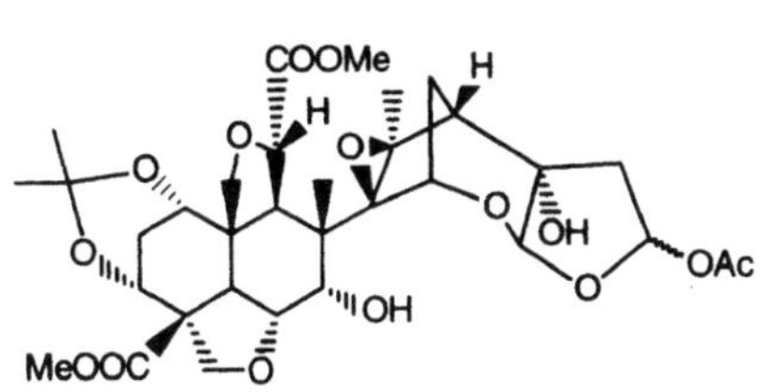

175

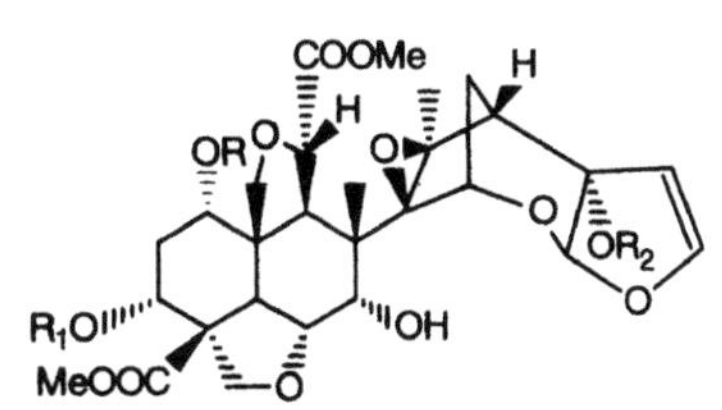

176 $R=R_1=R_2=Ac$
177 $R=R_2=H$, $R_1=Ac$
178 $R=H$, $R_1=R_2=Ac$

3-Tigloyazadirachtol (**119**) on exposure to triethylamine in aqueous methanol forms azadirachtol (**118**) which may be reesterified regioselectively with tigloyl chloride (*155*).

2.18.4.5. Functional Group Chemistry of Azadirachtol

Azadirachtol (**118**) is an ideal prototype for studying end game strategies concerned with azadirachtin total synthesis. The 1,3-diol of azadirachtol (**118**), 22,23-dihydroazadirachtol and 23-α/β-acetoxy-22, 23-dihydroazadirachtol (**174**) may be converted to an acetonide (**175**) to protect ring A (*156*). In the latter case, deprotection is facile using copper (II) chloride monohydrate in ethanol (*105*).

Treatment of azadirachtol (**118**) with acetic anhydride gives 1,3,20-triacetylazadirachtol (**176**) in high yield (*156*) whereas acetic pivalic anhydride (*66*) affords predominantly 3-acetylazadirachtol (**177**). From this result the reactivity pattern of an hypothetical azadirachtin derived pentaol may be extrapolated as $C(3)OH \sim C(11)OH > C(20)OH > C(1)OH \gg C(7)OH$. Therefore, to introduce a tiglate residue at C-1 will require an alternative approach.

2.18.4.6. Oxidation Reactions

Oxidation of azadirachtin using Cornforth's reagent (*59*) was not possible; however, slow oxidation of detigloylazadirachtin (**137**) to an A ring 1-ene-3-one was noted (*46*). The resistance of the C(7)OH group to oxidation may be due to shielding by the right-hand side of the molecule, which is held in place by a network of hydrogen bonding. Disruption of this hydrogen bonding, particularly the strong interaction between C(11)OH and the oxirane, might reduce the encumbrance of the C(7)OH by populating more open conformers such as that given for 7,11, 20-trimethoxyazadirachtin (**166**) (*304*). Activated DMSO reagent (*170, 301*) and ruthenium based oxidants (*96, 97*) failed to oxidize **179**. However, pyridinium dichromate (PDC) (*51, 57*) was successful in producing 23α-acetoxy-22,23-dihydro-7-keto-11-methoxyazadirachtin (**180**) which was transformed into 7-keto-11-methoxyazadirachtin (**181**) by pyrolytic *syn*-elimination of acetic acid.

PDC oxidation was successful for a various azadirachtin derivatives including 22,23-dihydro-11-methoxyazadirachtin (**184**) and 22,23-dihydro-11,20-methoxyazadirachtin (*158*) (*Vide infra*). Interestingly, azadirachtin on treatment with periodinane afforded allylic acetate (**182**) (*61*) instead of being oxidised. This complication was easily circumvented by masking the C(22)-C(23) enol ether as an anomeric acetate. Although capricious, oxidation was successful and gave 7-ketoazadirachtin following elimination of acetic acid (*158*).

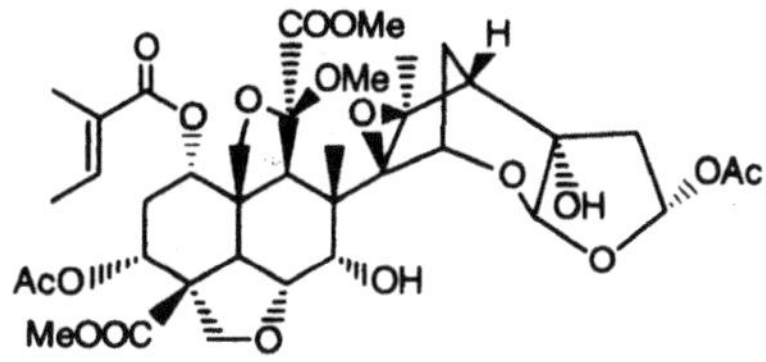

179

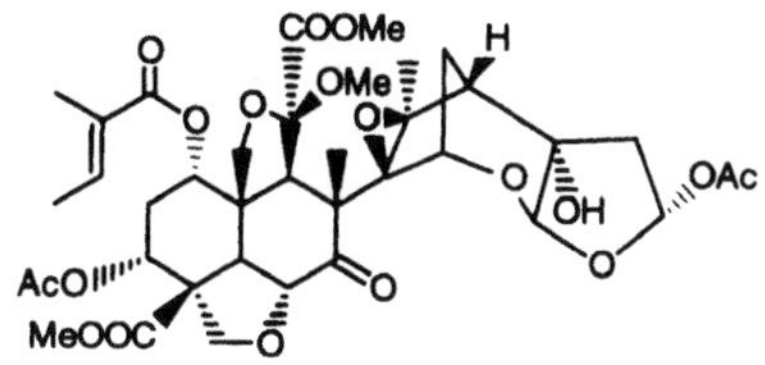

180

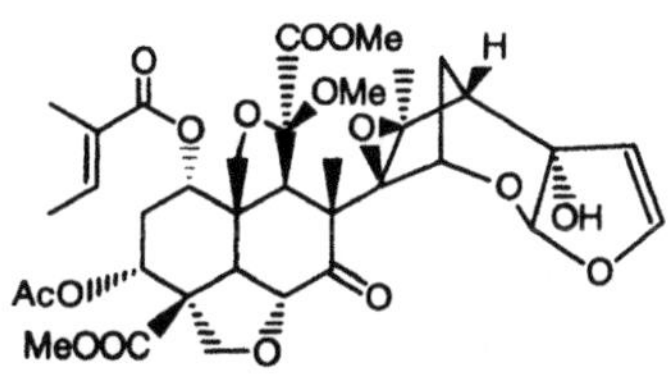

181

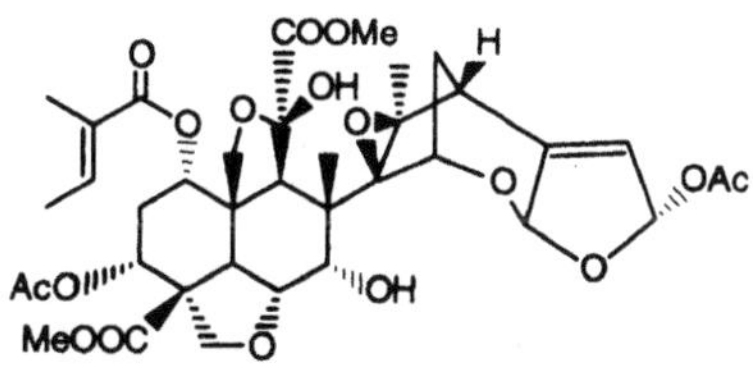

182

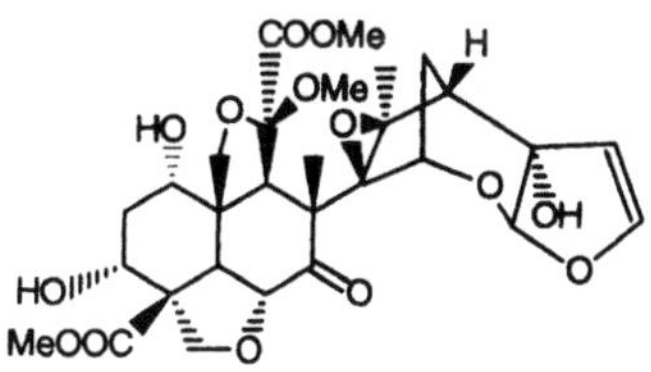

183

2.18.4.7. Functional Group Chemistry of 7-Keto Azadirachtins

Generally, 7-keto-azadirachtins behave in a fashion similar to their naturally configured C(7)OH counterparts. C(7)OH oxidation is accompanied by attenuation of C(1)OH reactivity, *e.g.* saponification of 7-keto-11-methoxyazadirachtin (**181**) occurs readily giving 3-desacetyl-1-detigloyl-7-keto-11-methoxyazadirachtin (**183**), whereas for 11-methoxyazadirachtin (**164**) the reaction is sluggish and incomplete even after 72 h. 1,3-Benzylidene acetal formation is possible for 7-ketoazadirachtins.

2.18.4.8. Retro-Aldol Reaction

C(8)-C(14) bond cleavage of azadirachtin was envisioned as part of synthetic programs aimed at making two functionalized components separately and then joining them. This might furnish fragment molecules which could be correlated with advanced synthetic material and used to study recombination for the purpose of total synthesis. One such cleavage to decalin and dihydrofuran fragment might involve a retroaldol reaction which would require C(7)OH oxidation and ring-opening of the tetra-substituted C(13)-C(14) oxirane. However base and acid mediated cleavage of the oxirane ring was unsuccessful.

On the other hand, oxidation of 22,23-dihydro-11-methoxyazadirachtin (**184**) with PDC gave **185** and **186** (*158*, *161*, *162*). The unusual cyclic carbonate (**186**) possess the necessary disposition of functional groups to allow intramolecular oxirane ring-opening by a series of β-elimination reactions. Additionally, the boat configured B ring fulfills the Corey-Sneen (*55*) requirements for C(8)-C(14) bond cleavage by placing this bond orthogonal to the C-7 carbonyl residue.

Thus treatment of ketocarbonate (**186**) with sodium methoxide in methanol caused a retro-aldol reaction to take place giving decalin (**187**) as a mixture of C-8 epimers (*161*, *162*) and the unusual spirocycle (**188**) (*161*). The formation of **186** was optimized by using the more acidic oxidising agent pyridinium chlorochromate (PCC) (*56*, *58*). Protection of the 1,3-diol unit of **187** as a diacetate (**189**), benzyledene acetal (**190**), or acetonide (**191**) was readily accomplished by standard means (*162*). Similarly, oxidation of 11-acetoxy-22,23-dihydroazadir-

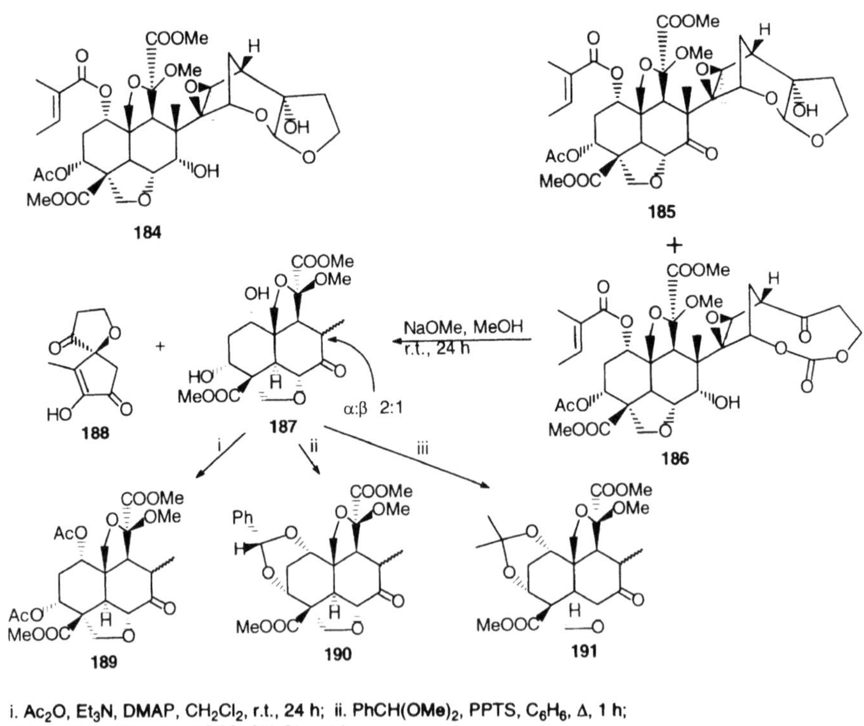

i. Ac_2O, Et_3N, DMAP, CH_2Cl_2, r.t., 24 h; ii. $PhCH(OMe)_2$, PPTS, C_6H_6, Δ, 1 h; iii. 2-methoxypropene, PPTS, CH_2Cl_2, 24 h

achtin (**192**) with PCC gave the required ketocarbonate (**194**) in reduced yield *(161)*, the retro-aldol reaction of which was successful and accompanied by C(11)OAc hydrolysis. The resulting decalins (**195**, **196**, **197**) were then readily transformed into the fully protected fragments (**198**) and (**199**) which showed considerable homology with synthetic material.

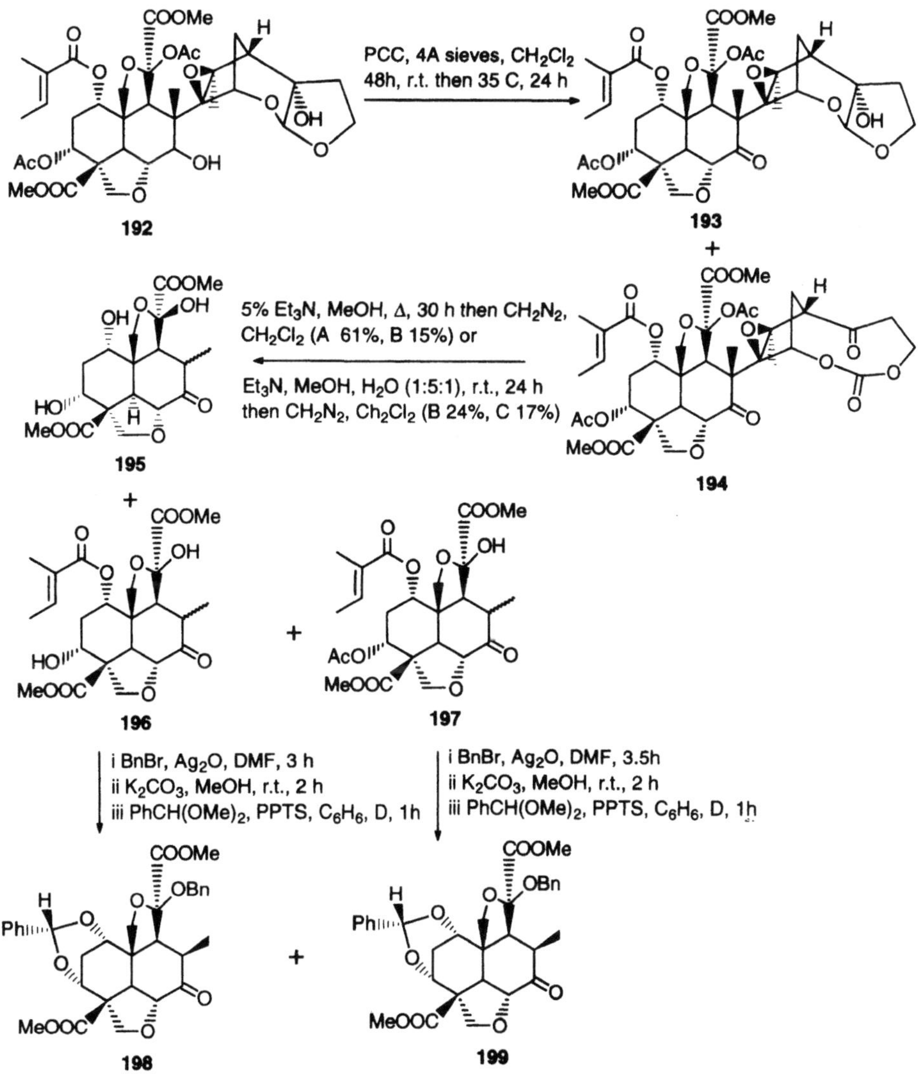

2.18.4.9. Skeletal Rearrangements

Rearrangement of azadirachtin to the azadirachtinins (**133**, **134**) was observed during an attempt at deoxygenation (*249*); this rearrangement was later induced under acidic conditions for both azadirachtin (**135**) and 3-tigloylazadirachtol derivatives (*36*, *156*). In the conversion of azadirachtin to 3-acetyl-1-tigloylazadirachtinol (**133**) C(22)-C(33) enol ether cleavage competes.

The rather unusual polycyclic oxetane (**200**) was isolated as a by-product of azadirachtin saponification. This was rationalized as follows. Oxetane ring formation by *7-endo-tet* cyclisation (*23*) induces opening of the C(13)-C(14) oxirane ring. The resulting C(14)OH group then causes transacetalization at C-11, thus liberating an angular C-19 hydroxymethylene 'arm' which lactonizes at the C-29 ester.

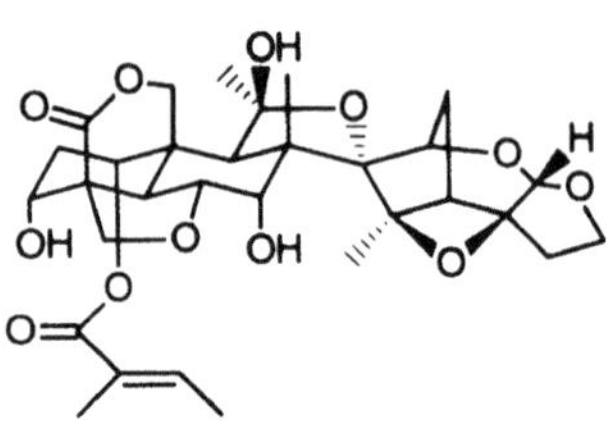

200

A series of cage-like C-7 acetals has been isolated from an attempt at acid catalysed retro-aldol decomposition of 7-ketoazadirachtins.

2.18.5. Synthesis

Since azadirachtin contains a plethora of oxygen functionalities of various nature, sixteen chiral centres seven of which are quaternary, and is both acid as well as base labile, its synthesis presents a formidable challenge. The structure has a left hand side multifunctional decalin and a right hand side polycyclic hydroxydihydrofuran fragment joined at C(8)-C(14).

Over the past few years LEY and cowokers (*5–7*, *39*, *98*, *99*, *116*, *117*, *165*) and others (*178*, *199*) have been engaged in intensive studies toward the total synthesis of azadirachtin, which is referred in Sections 2.18.5.1 and 2.18.5.2, namely preparation of the hydroxyfuran acetal fragment '**A**' and the decalin fragment '**B**' (*116*) and then formation of the C(8)-C(14) bond between the two fragments.

2.18.5.1. Synthesis of Dihydrofuranacetal Fragment 'A' (**213**)

Synthesis of **201** was achieved in 28 steps starting from enantiomerically pure aldehyde (**202**) which is readily available from *S*-(-)-ethyl lactate (*103*). Racemic dihydrofuranacetals (**203** and **204**) related to azadirachtin were available from the known lactone (**205**) (*171*). However, syntheses of the model compounds (**203** and **204**) afford no opportunity for introducing oxygen functionality at C-9 or C-10 which is needed if suitable fragments were to be made available for coupling studies. To fulfill this requirement, an alternative route was adopted by which bromoketone (**206**) was converted to the enantiomerically pure pivotal intermediate (**207**) in 12 steps (*6, 7, 164*).

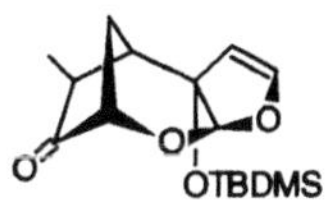

201

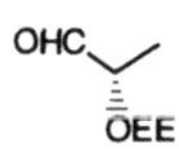

202

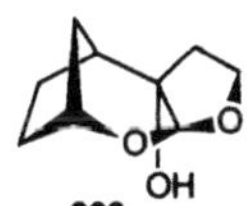

203

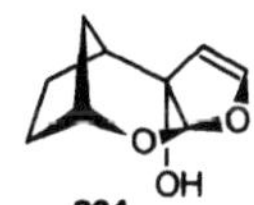

204

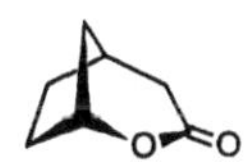

205

206

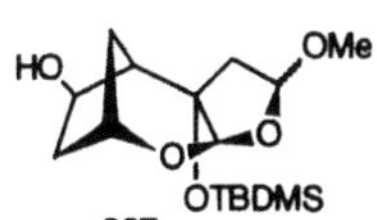

207

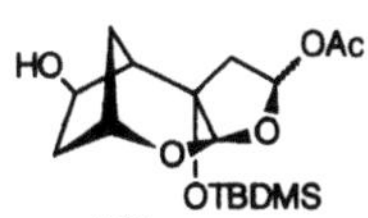

208

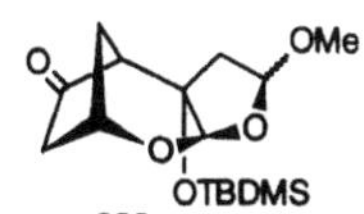

209

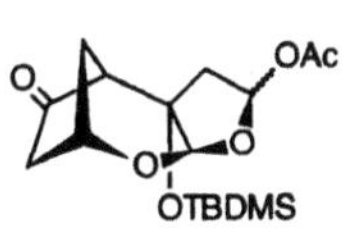

210

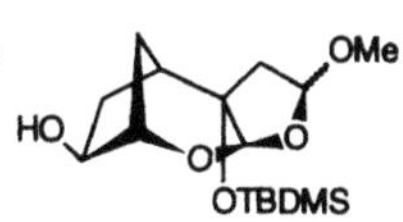

211

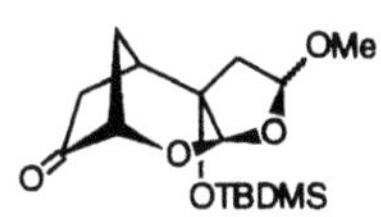

212

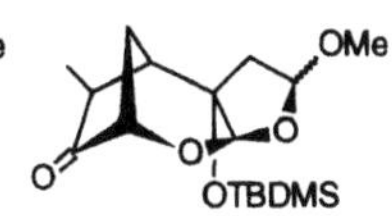

213

2.18.5.1.1. Preparation of Prototype Coupling Fragment

A flexible synthetic programme which gives access to a number of potential coupling intermediates from a common precursor was needed for C(8)-C(14) bond formation. Intermediate (**207**) satisfied this requirement and was readily transformed into a range of prototype compounds for coupling. Oxidation of **207** and **208** proceeded well using PDC (*51*) to give the corresponding C-9 ketones (**209** and **210**) in which the C-10 position is activated for coupling. Oxidation of **211** produced a ketone **212** which was stereoselectively monoalkylated to give the potential coupling fragment (**213**) (*7*).

2.18.5.2. Decalin 'B' (**221**) Synthesis

The synthesis of the decalin fragment requires the introduction of a number of contiguous stereogenic centres, several of which are fully substituted and whose stereochemistry must consequently be established with the correct relative configuration since no potential for equilibration exists.

MORI and WATANABE's (*178*) approach to the problem differed in that it involved formation of the C(8)-C(14) bond at a relatively early stage while SHIBASAKI *et al.*'s method (*241–243*) method for synthesizing the decalin fragment based on an asymmetric variant of the Heck reaction would have involved formation of C(8)-C(14) bond at the later stage. LEY and co-workers (*165*) prepared functionalised decalin (**216**) by an intramolecular *Diels-Alder* reaction of **214** and subsequent internal Michael addition to construct *trans*-decalin (**215**) which on stereoselective functional group interconversions followed by 1,3-diol protection afforded **216**. The remaining problem was oxidation of the potential C-19 methyl in order to construct C-9, C-10 tetrahydrofuranacetal moiety of azadirachtin; however all efforts in this direction failed. To avoid this difficulty, **217** with a dimethyl(phenyl)silyl group as a stereocontrol element to increase was subjected to the *Diels-Alder* reaction to afford **218** in 3 steps (*159*). The latter was converted to **221** by way of **219** (*246*) and **220**.

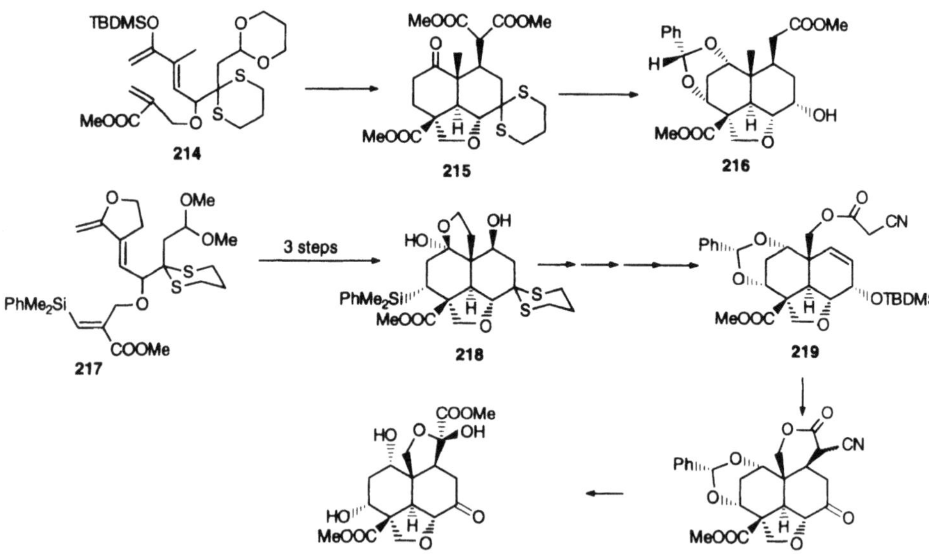

2.18.5.3. Coupling of 'A' and 'B' Fragments

At all this stage, syntheses of the right hand side dihydrofuranacetal (**213**) and the left hand side fully functionalised decalin (**221**) fragments had been achieved separately. Coupling of these two units to afford the desired C(8)-C(14) bond linkage would complete a total synthesis.

3. Other Compounds

3.1. Diterpenoids

Several tricyclic diterpenoids have been isolated from root and stem bark of neem; their structures are shown in Scheme 7. Root bark has yielded sugiol, nimbiol (*247*), morgocin, morgocilin, morgocinin (*18*), nimbidiol (*169*), nimbilicin, nimbocidin (*11*) and nimolinin (*13*) whereas stem bark has provided nimbisonol, nimbosodione, demethyl nimbionol (*16*), nimbinone, nimbione (*8*), nimosone, methyl nimbiol, methyl nimbionone, nimbosone (*9*), nimbionol, nimbionone (*252*), nimbinone, nimbonolone (*10*), morgosone, morgosolone (*17*), morgolone, morgolonone and isomorgolonone (*14*).

3.2. Steroids and Other Triterpenoids

Cycloeucalenol, 24-methylenecycloartanol (*74*), β-sitosterol and its β-D-glucoside (*40*), 4,14α-dimethyl-5α-ergosta-8,24(28)-dien-3β-ol, 4α-methyl-5α-ergosta-8,24(28)-dien-3β-ol (*24*) and 24-methylene lophenol (*25*) have been isolated from heartwood while cholesterol (*257*) and stigmasterol (*264*) have been reported in flowers. Two new glycosides of stigmasterol have been isolated from seeds and α- and β-nimolactones have been isolated from fresh fruit coatings (*260*). These compounds belong to the new class of enneanortriterpenoids lacking the entire C-17 side chain and one carbon atom in ring D. Limbonin, a trinortriterpenoid-D-lactone with open C-17 side chain and oxidised C-29, was isolated from the neutral fraction of neem kernel extract (*262*) (Scheme 8).

3.3. Phenolic Compounds

3.3.1. Flavonoids

Quercetin and isorhamnetin have been reported from flowers and leaves (*28*, *263*) but kaempferol and myricetin have been reported in flowers only (*299*). Melicitrin, a new myricetin glycoside along with kaempferol-3-glucoside and quercetin-3-galactoside was isolated from flowers (*291*). An isoprenylated flavone, nimbaflavone, has been reported

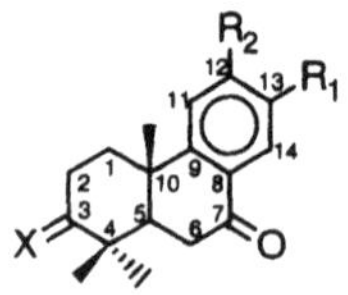

Sugiol R_1=Me, R_2= OH, X=H,H

Nimbisonol R_1=Me, R_2=OH, X=β–OH,H

Nimbinone R_1=Me, R_2=OH, X=O

Nimbiol R_1=^{i}Pr, R_2=OH, X=H,H

Nimosone R_1=^{i}Pr, R_2=OH, X=O

Methyl nimbiol R_1=^{i}Pr, R_2=OMe, X=H,H

Margocin R_1=^{i}Pr, R_2=H, X=O

Margocilin R_1=^{i}Pr, R_2=OH, X=β-OH,H

Margocinin R_1=$MeCHCH_2OH$, R_2=OH, X=O

Nimbidiol R_1=R_2=OH, X=H,H

Nimbionol R_1=OMe, R_2=OH, X=β-OH,H

Nimbionone R_1=OMe, R_2=OH, X=O

Demethylnimbionol R_1=R_2=OH, X=β-OH,H

Methylnimbionone R_1=R_2=OMe, X=O

Nimbione R_1=OH, R_2=Me, X=O

Nimbonone R_1=OMe, R_2=Et, X=H,H

Nimbosodione R_1=Ac, R_2=OH, X=H,H

Margolone R_1=COOH, R_2=Me, X=H,H

Margolonone R_1=COOH, R_2=Me, X=O

iso-Margolonone R_1=Me, R_2=COOH, X=O

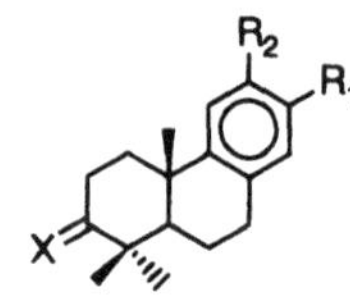

Nimbosone R_1=Ac, R_2=OMe, X=H,H

Nimbonolone R_1=OMe, R_2=Et, X=O

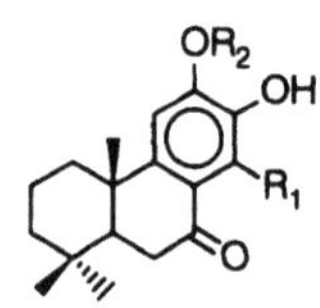

Margosone R_1=H, R_2=^{i}Pr

Margosolone R_1=Me, R_2=H

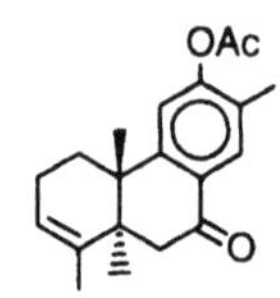

Nimbilicin

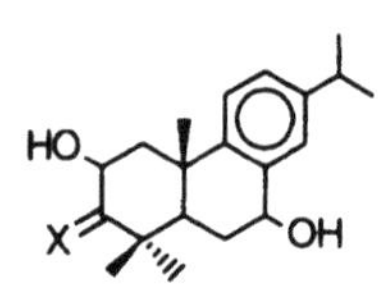

Nimbocidin

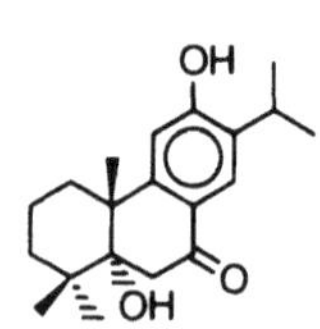

Nimolinin

Scheme 7. Tricyclic diterpenoids present in various parts of neem tree

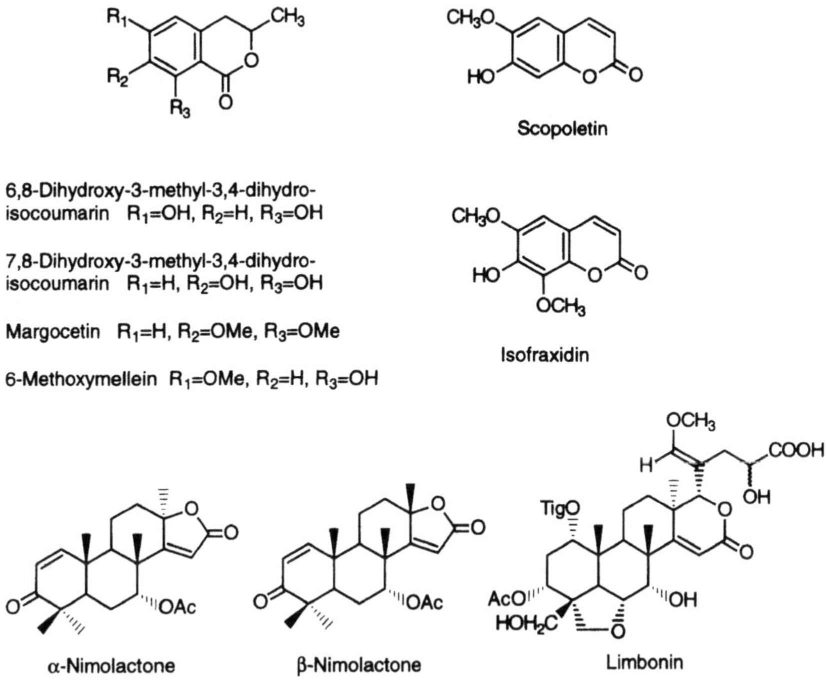

Scheme 8. Some of the coumarins, triterpenes and steroids found in neem

in the leaves (*84*) (Scheme 9). This is probably the first isoprenyl-flavanone from Meliaceae.

3.3.2. Flavonoglycosides

Two glycosides have been reported from flowers and one from the leaves: Quercetin-3-galactoside and kaempferol-3-glucoside from the flowers and myrcetin-3′-larabinoside from the leaves (*291*).

3.3.3. Coumarins

The acidic fraction of a CH_2Cl_2 extract of fresh, uncrushed winter twigs yielded scopoletin, 6,8-dihydroxy-3-methyl-3,4-dihydroisocoumarin and 7,8,-dihydroxy-3-methyl-3,4,-dihydroisocoumarin while the acidic fraction of an EtOH extract contained margocetin, 6-methoxymellein and isofraxidin along with scopoletin (*268*) (Scheme 8). Scopoletin was also isolated from the leaves (*272*).

3.3.4. Dihydrochalcone

The aqueous fraction of the fruits yielded nimbochalcin, a dihydrochalcone derivative (*263*) (Scheme 9).

3.3.5. Tannins

Gallic acid, (+)-gallocatechin, (−)-epicatechin, (+)-catechin and epigallocatechin have been isolated from the aqueous extract of bark (*311*).

3.4. Carbohydrates and Proteins

The gum exudate from the stem of old neem trees is a mixture of proteins and polysugars. Sugars and proteins are tightly interlinked. Due to this complexity the structure elucidation of proteins and polysaccharides was a complex job (*3*). D-glucose, D-glucoronic acid, L-arabinose, L-fucose (*179*), mannose, xylose, rhamnose (*3*), D-glucosamine (*142*), aldobiouronic acid (*21*), 4-*O*-(4-*O*-methyl-α-D-glucopyranosyl uronic acid)-D-galactose and aldotriouronic acids (*21*) were reported. Investigations on the amino acid component of the gum have also been carried out (*3*, *4*).

3.5. Sulphur Compounds

The steam volatile fraction of neem oil was reported to be contraceptive (*224*). Several tri- and tetrasulfides have been identified from the steam volatile fraction of the matured leaves by GC-MS analysis (*198*). GC-MS analysis of volatile fraction of the crushed seeds also yielded di- and trisulphides having di-*n*-propyl disulphide (75.74%) as the major component (*22*) (Scheme 9).

3.6. Hydrocarbons, Acids and Esters

The long chain hydrocarbon fraction obtained from leaves (*20*, *48*) was reported to consist of a mixture of octadecane, nonadecane, hexacosane, nonacosane, tetratriacontane, *n*-hexacosanol, β-carotene and xanthophyll while in fruit coats, icosane, docosane, 2-methyltricosane and docasene were identified (*258*). A substituted aromatic ester,

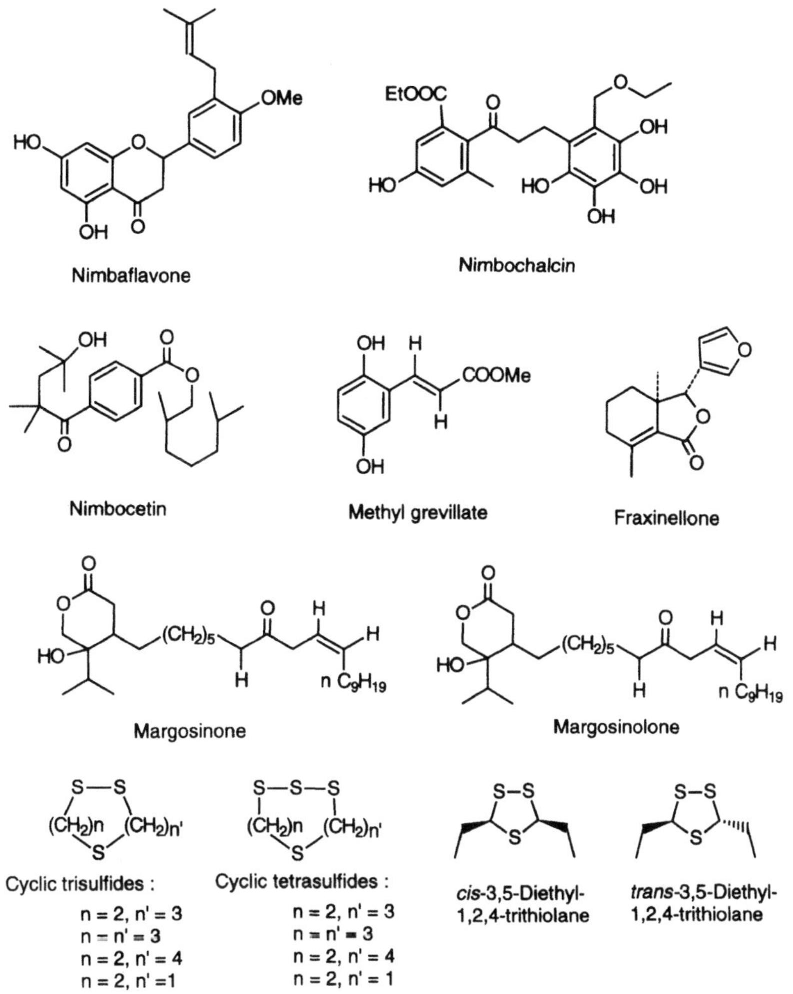

Scheme 9. Flavores, dihydrochalcones, hydroxymethylfurfurals and sulphur compounds found in different parts of neem tree

nimbocetin, was isolated from fruits (*263*), and methyl grevillate (*10*), margosinone and margosinolone, two new polyacetate derivatives (*12*), were reported in stem bark (Scheme 9). Oxalic acid was reported in leaves and indole acetic acid, indole pyruvic acid, tiglic acid and fatty acids from seeds (*299*). Neem kernels contain approximately 22–45% fixed oil having glycerides of oleic (53%), stearic (19%), palmitic (16%), linoleic (11%) alongwith minor amounts of arachidic, behenic, lignoceric and myristic acids (*63*, *232*). 5-Hydroxymethylfurfural was isolated from the fruits (*263*). Fraxinellone, a degraded terpene was found in bark (*71*) (Scheme 9).

References

1. AKHILA, A., M. SRIVASTAV, and K. RANI: Production of Radioactive Azadirachtin in the Seed Kernels of *Azadirachta indica*. Nat. Prod. Letters, **11**, 107 (1998).
2. ALLEN, E.J., J.P. EASWARA, S. JOHNSON, J. MORDUE (Luntz), E.D. MORGAN, and T. STUCHBURY: The Production of Azadirachtin by *in-vitro* Tissue Cultures of Neem, *Azadirachta indica*. Pest. Sci., **42**, 147 (1994).
3. ANDERSON, D.M.W., and A. HENDRIE: The Proteinaceous Gum Polysaccharide from *Azadirachta indica* A. JUSS. Carbohydrate Research, **20**, 259 (1971).
4. ANDERSON, D.M.W., A. HENDRIE, and A.C. MUNRO: The Amino Acid and Amino Sugar Composition of Some Plant Gums. Phytochem., **11**, 733 (1972).
5. ANDERSON, J.C., and S.V. LEY: Chemistry of Insect Antifeedants from *Azadirachta indica* (Part 6): Synthesis of an Optically Pure Acetal Intermediate for Potential Use in the Synthesis of Azadirachtin and Novel Antifeedants. Tetrahedron Letters, **31**, 431 (1990).
6. ANDERSON, J.C., and S.V. LEY: Chemistry of Insect Antifeedants from *Azadirachta indica* (Part 7): Preparation of an Optically Pure Hydroxy Acetal Epoxide Related to Azadirachtin. Tetrahedron Letters, **31**, 3437 (1990).
7. ANDERSON, J.C., S.V. LEY, D. SANTAFIANOS, and R.N. SHEPPARD: Chemistry of Insect Antifeedant from *Azadirachta indica* (Part 8): Synthesis of Hydroxy Dihydrofuran Acetal Fragments for Biological Evaluation and Azadirachtin Total Synthesis Studies. Tetrahedron, **47**, 6813 (1991).
8. ARA, I., B.S. SIDDIQUI, S. FAIZI, and S. SIDDIQUI: Terpenoids from the Stem Bark of *Azadirachta indica*. Phytochem., **27**, 1801 (1988).
9. ARA, I., B.S. SIDDIQUI, S. FAIZI, and S. SIDDIQUI: Tricyclic Diterpenoids from the Stem Bark of *Azadirachta indica*. J. Nat. Prod., **51**, 1054 (1988).
10. ARA, I., B.S. SIDDIQUI, S. FAIZI, and S. SIDDIQUI: Diterpenoids from the Root Bark of *Azadirachta indica*. Phytochem., **28**, 1177 (1989).
11. ARA, I., B.S. SIDDIQUI, S. FAIZI, and S. SIDDIQUI: Diterpenoids from the Root Bark of *Azadirachta indica*. Z. Naturforsch. B. Chem. Sci., **44**, 1279 (1989).
12. ARA, I., B.S. SIDDIQUI, S. FAIZI, and S. SIDDIQUI: Margosinone and Margosinolone, Two New Polyacetate Derivatives from *Azadirachta indica*. Fitoterapia, **60**, 519 (1989).
13. ARA, I., B.S. SIDDIQUI, S. FAIZI, and S. SIDDIQUI: Two new terpenoids from Root Bark of *Azadirachta indica*. J. Nat. Prod., **52**, 1209 (1989).
14. ARA, I., B.S. SIDDIQUI, S. FAIZI, and S. SIDDIQUI: Structurally Novel Diterpenoid Constituents from the Stem Bark of *Azadirachta indica* (Meliaceae). J. Chem. Soc., Perkin Trans. I, 343 (1989).
15. ARA, I., B.S. SIDDIQUI, S. FAIZI, and S. SIDDIQUI: Isolation of Meliacin Cinnamates from the Root Bark of *Azadirachta indica* A. JUSS (Meliaceae). Heterocycles, **29**, 729 (1989).
16. ARA, I., B.S. SIDDIQUI, and S. FAIZI: New Diterpenoids from the Stem Bark of *Azadirachta indica*. J. Nat. Prod., **53**, 816 (1990).
17. ARA, I., B.S. SIDDIQUI, S. FAIZI, and S. SIDDIQUI: Tricyclic Diterpenes from the Stem Bark of *Azadirachta indica*. Planta Medica, **56**, 84 (1990).

18. ARA, I., B.S. SIDDIQUI, S. FAIZI, and S. SIDDIQUI: Tricyclic Diterpenoids from the Root Bark of *Azadirachta indica*. Phytochem., **29**, 911 (1990).
19. ARA, I., B.S. SIDDIQUI, S. FAIZI, and S. SIDDIQUI: Isolation and Structure Elucidation of the Triterpene Azadirinin from the Roots of *Azadirachta indica*. Fitoterapia, **63**, 118 (1992).
20. AWASTHI, Y.C., and C.R. MITRA: Constituents of *Melia indica* leaves. Phytochem., **10**, 2842 (1971).
21. BAJPAI, K.S., V. CHANDRASEKHARAN, S. MUKHERJEE, and A.N. SRIVASTAVA: Isolation of Pure Aldobiouronic Acids and Aldotriouronic Acids from Plant Gums and Mucilages. Indian J. Chem., **8**, 48 (1970).
22. BALANDRIN, M.F., S.M. LEE, and J.A. KLOCKE: Biologically Active Volatile Organosulfur Compounds from Seeds of the Neem Tree, *Azadirachta indica* (Meliaceae). J. Agric. Food Chem., **36**, 1048 (1988).
23. BALDWIN, J.E.: Approach Vector Analysis – A Stereochemical Approach to Reactivity. J. Chem. Soc., Chem. Commun., 738 (1976).
24. BANERJI, R., G. MISRA, and S.K. NIGAM: On the Triterpenes of *Azadirachta indica* (*Melia azadirachta*) Fitoterapia, **48**, 166 (1977).
25. BANERJI, R., G. MISRA, and S.K. NIGAM: Identification of 24-Methylene Lophenol from Heartwood of *Azadirachta indica*. Phytochem., **26**, 2644 (1987).
26. BANERJI, B., and S.K. NIGAM: Wood Constituents of Meliaceae: A Review. Fitoterapia, **55**(1), 3 (1984).
27. BANIK, B.K., S. GHOSH, and U.R. GHATAK: Stereoselective Synthesis of (+)-Nimbidiol. Indian J. Chem., **27**B, 103 (1988).
28. BASAK, S.P., and D.P. CHAKRABORTY: Chemical Investigation of *Azadirachta indica* Leaf (*M. azadirachta*). J. Indian Chem. Soc., **45**, 466 (1968).
29. BAX, A., and G.A. MORRIS: An Improved Method for Heteronuclear Chemical Shift Correlation by Two-dimensional NMR. J. Magn. Reson., **42**, 501 (1981)
30. BENDALL, M.R., D.M. DODDRELL, and D.T. PEGG: Editing of ^{13}C NMR Spectra. A Pulse Sequence for the Generation of Subspectra. J. Am. Chem. Soc., **103**, 4603 (1981).
31. BHARGAVA, K.P., M.B. GUPTA, G.P. GUPTA, and C.R. MITRA: Anti-inflammatory Activity of Saponins and Other Natural Products. Indian J. Med. Res., **58**, 724 (1970).
32. BHOWMICK B.N., and B.K. CHOUDHARY: Antifungal Activity of Leaf Extracts of Medicinal Plants on *Alternaria alternata*. Indian Botanical Reporter, **1**, 164 (1982).
33. BHIDE, N.K., D.J. MEHTA, and H.A. LEWIS: Diuretic Action of Sodium nimbidinate. Ind. J. Med. Sci., **12**, 141 (1958).
34. BILTON, J.N., H.B. BROUGHTON, S.V. LEY, Z. LIDERT, E.D. MORGAN, H.S. RZEPA, and R.N. SHEPPARD: Structural Reappraisal of the Limonoid Insect Antifeedant, Azadirachtin. J. Chem. Soc., Chem. Commun., 968 (1985).
35. BILTON, J.N., H.B. BROUGHTON, P.S. JONES, S.V. LEY, Z. LIDERT, E.D. MORGAN, H.S. RZEPA, R.H. SHEPPARD, A.M.Z. SLAWIN, and D.J. WILLIAMS: An X-ray Crystallographic, Mass Spectroscopic and NMR Study of the Limonoid Insect Antifeedant, Azadirachtin and Related Derivatives. Tetrahedron, **43**, 2805 (1987).
36. BILTON, J.N., P.S. JONES, and S.V. LEY: Chemistry of Insect Antifeedants from *Azadirachta indica* (Part 1): Conversion from the Azadirachtin to the Azadirachtinin Skeleton. Tetrahedron Letters, **29**, 1849 (1988).
37. BLANEY, W.M., M.S.J. SIMMONDS, S.V. LEY, J.C. ANDERSON, S.C. SMITH, and A. WOOD: Effect of Azadirachtin-Derived Decalin (Perhydronaphthalene) and Dihydrofuranacetal (Furo[2, 3-b] Pyran) Fragments on the Feeding Behaviour of *Spodoptera littoralis*. Pest. Sci., **40**, 169 (1994).

38. BOKEL, M., R. CRAMER, H. GUTZEIT, S. REEB, and W. KRAUS: Tetranortriterpenoids Related to Nimbin and Nimbolide from *Azadirachta indica* A. JUSS. (Meliaceae). Tetrahedron, **46**, 775 (1990).
39. BRASCA, M.G., H.B. BROUGHTON, D. CRAIG, S.V. LEY, A.A. SOMOVILLA, and P.L. TOOGOOD: Chemistry of Insect Antifeedants from *Azadirachta indica* (Part 2): Synthesis of Polyoxygenated Decalin with Limonoid Structural Homology. Tetrahedron Letters, **29**, 1853 (1988).
40. BROUGHTON, H.B., S.V. LEY, A.M.Z. SLAWIN, D.J. WILLIAMS, and E.D. MORGAN: X-ray Crytallographic Structure Determination of Detigloyldihydroazadirachtin and Reassignment of the Structure of the Limonoid Insect Antifeedant, Azadirachtin. J. Chem. Soc., Chem. Commun., 46 (1986).
41. BRUHN, A., M. BOKEL, and W. KRAUS: 4α, 6α-Dihydroxy-A-homoazadiron, ein neues Tetranorterpenoid aus *Azadirachta indica* A. JUSS (Meliaceae). Tetrahedron Letters, **25**, 3691 (1984).
42. BUCHANAN, J.G.ST.C., and T.G. HALSALL: Conversion of a Simple Meliacin (7α-Acetoxymeliaca-14,20,22-trien-3-one) into Azadirone and of Khayanthone into Khivorin. Chem. Commun., 1493 (1969).
43. BUCHANAN, J.G.ST.C., and T.G. HALSALL: The Conversion of Turreanthin and Turreanthin A into Simple Meliacins by a Route Involving an Oxidative Rearrangement of Probable Biogenetic Importance. J. Chem. Soc. (C), 2280 (1970).
44. BUTTERWORTH, J.H., and E.D. MORGAN: Isolation of a Substance that Suppresses Feeding in Locusts. J. Chem. Soc., Chem. Commun., 23 (1968).
45. BUTTERWORTH, J.H., and E.D. MORGAN: Locust Feeding Inhibition of the Seeds of the Neem Tree, *Azadirachta indica*. J. Insect Physiol. **17**, 969 (1971).
46. BUTTERWORTH, J.H., E.D. MORGAN, and G.R. PERCY: Structure of Azadirachtin Functional groups. J. Chem. Soc., Perkin Trans. I, (19), 2445 (1972).
47. CHAMPAGNE, D.E., O. KOUL, M.B. ISMAN, G.G.E. SCUDDER, and G.H.N. TOWERS: Biological Activity of Limonoids from the Rutales. Phytochem., **31**, 377 (1992).
48. CHAVAN, S.R.S.: Chemistry of Alkanes Separated from Leaves of *Azadirachta indica* and Their Larvicidal/Insecticidal Activity against Mosquitoes. In: Natural Pesticides from the Neem Tree and other Tropical Plants (H. Schmutterer and K.R.S. Ascher, eds.), p. 59. Eschborn, FRG: GTZ. 1984. Proc. 2nd Int. Neem Conf. Rauischholzhausen. 1983.
49. CHAWLA, A.S., M. KUMAR, and I. BANSAL: Chemical Constituents and Biological Activity of 'Neem' – A Review. Indian Drugs, **32**, 57 (1995).
50. CHEN, S-H., R.F. HORVATH, J. JOGLER, M. J. FISHER, and S.J. DAMISHEFSKY: Application of the Ibuka-Yamamoto Reaction to a Problem in Stereochemical Communication – A Strategy for the Stereospecific Synthesis and Stabilization of the Triene Substructure of Rapamycin through Sulfone Substitution. J. Organ. Chem., **56**, 5834 (1991)
51. COATES, W.M., and J.R. CORRIGAN: Pyridine Dichromate as an Oxidising Agent. Chem. and Ind., 1594 (1969).
52. CONNOLLY, J.D.: Chemistry of the Limonoids of the Meliaceae and the Cneoraceae. In: Chemistry and Chemical Taxonomy of the Rutales (P.G. Waterman and M.F. Grundon, eds.), p. 175. New York: Academic Press. 1983.
53. CONOLLY, J.D., K.L. HANDA, and R. MCCRINDLE: Further Constituents of Nim Oil: The Constitution of Meldenin. Tetrahedron Letters, 437 (1968).
54. CONOLLY, J.D., K.H. OVERTON, and J. POLONSKY: The Chemistry and Biochemistry of the Limonoids and Quassinoids. In: Progress in Phytochemistry **2** (L. REINHOLD

and Y. LIWSCHITZ, eds), p. 385. London: Interscience Publisher, J.Wiley & Sons. 1970.

55. COREY, E.J., and R.W. HAHL: Synthesis of Limonoid, Azadiradione. Tetrahedron Letters, **30**, 3023 (1989).

56. COREY, E.J., and G. SCHMIDT: Useful Procedures for the Oxidation of Alcohols involving Pyridinium Dichromate in Aprotic Media. Tetrahedron Letters, **20**, 399 (1979).

57. COREY, E.J., and R.A. SNEEN: Stereoelectronic Control in Enolization-Ketonization Reactions. J. Amer. Chem. Soc., **78**, 6269 (1956).

58. COREY, E.J., and J.W. SUGGS: Pyridinium dichromate – An Efficient Reagent for Oxidation of Primary and Secondary Alcohols to Carbonyl Compounds. Tetrahedron Letters, **16**, 2647 (1975).

59. CORNFORTH, R.H., J.W. CORNFORTH, and G. POPJAK: Preparation of (R)- and (S)-Mevalonolactones. Tetrahedron, **18**, 1351 (1962)

60. DESHPANDE, V.Y., K.N. MENDULKAR, and N.L. SADRE: Male Antifertility Activity of *A. indica* in Mice, a Preliminary Report. J. Postgraduate Medicine, **26**, 167 (1980).

61. DESS, D.B., and J.C. MARTIN: A Useful 12-I-5 Triacetoxyperiodinane (the Dess-Martin Periodinane for the Selective Oxidation or Secondary Alcohols and a Variety of Related 12-I-5 Species. J. Amer. Chem. Soc., **113**, 7277 (1991).

62. DEVAKUMAR, C., and B.K. GOSWAMI: Nematicidal Principles from Neem. III. Isolation and Screening of Neem Meliacins. Pesticide Res. J., **4**, 79 (1992).

63. DEVAKUMAR, C., and S.K. MUKHERJEE: Chemistry of Neem Bitter Principles. I.A.R.I. Res. Bull., **40**, 1 (1983).

64. DEVAKUMAR, C., and S.K. MUKHERJEE: 4-Epinimbin, a New Meliacin from *Azadirachta indica* A. JUSS. Indian J. Chem., **24**B, 1105 (1985).

65. DEVAKUMAR, C., and S.K. MUKHERJEE: Chemical and Spectral Studies on Nimbin Group of Neem (*Azadirachta indica*) Constituents. Pesticide Res. J., **4**, 87 (1992).

66. D'INCAN, E., S. SIBILLE, and J. PERICHON: Electrosynthesis of Ketones from Organic Halides and Anhydrides. Tetrahedron Letters, **27**, 4175 (1986).

67. DIXIT, V.P., R. SINHA, and R. TANK: Effect of Neem Seed Oil on the Blood Glucose Concentration of Normal and Alloxan Diabetic Rats. J. Ethnopharmacol., **17**, 95 (1986).

68. DREYER, D.L.: Biogenetic Relationships of Degraded Triterpenes in the Rutales. In: Isoprenoids in Plants: Biochemistry and Function (W.D. NES, G. FULLER and L.-S. TSAI, eds.), p. 247. New York: Marcel Dekker. 1984.

69. EKANEM, O.J.: Has *Azadirachta indica* (Dongoyaro) any Antimalarial Activity? Nigerian Medical Journal, **8**, 8 (1978).

70. EKONG, D.E.U.: Chemistry of the Meliacins (Limonoids): The Structure of Nimbolide, a New Meliacin from *Azadirachta indica*. J. Chem. Soc., Chem. Commun., 808 (1967).

71. EKONG, D.E.U., C.O. FACUNLE, A.K. FASINA, and J.I. OKOGUN: The Meliacins (Limonoids). Nimbolin A and B – Two New Meliacin Cinnamates from *Azadirachta indica* L. and *Melia Azedarach*. L. J. Chem. Soc., Chem. Commun., 1166 (1969).

72. EKONG, D.E.U., and S.A. IBIYEMI: Biosynthesis of Nimbolide from [2-^{14}C,(4R)4-$^{3}H_{1}$] Mevalonic Acid Lactone in the Leaves of *Azadirachta indica*. Phytochem., **24**, 2259 (1985).

73. EKONG, D.E.U., S.A. IBIYEMI, and E.O. OLAGBEMI: The Meliacins (Limonoids) – Biosynthesis of Nimbolide in the Leaves of *Azadirachta indica*. J. Chem. Soc., Chem. Commun., 1117 (1971).

74. EKONG, D.E.U., E.O. OLAGBEMI, and A.I. SPIFF: Cycloeucalenol and 24-Methylenecycloartanol in Wood Oils from the Family Meliaceae. Chem. and Ind., 1808 (1968).
75. EKUNDAYO, O.: Biosynthesis of Nimbolide in *Azadirachta indica* A. JUSS from (2-^{14}C)-Mevalonate and (2-^{14}C)-Acetate. Z. Pflanzenphysiol. Bd. **112**, 139 (1983).
76. FREEMAN, R., and G.A. MORRIS: Experimental Chemical Shift Correlation Maps in Nuclear Magnetic Resonance Spectroscopy. J. Chem. Soc., Chem. Commun., 684 (1978).
77. FUJIWARA, T., T. TAKEDA, Y. OGIHARA, M. SCHIMUZU, T. NOMURA, and Y. TOMITA: Studies on the Structures of Polysaccharides from the Bark of *Melia azadirachta*. Chem. Pharm. Bull., **30**, 4025 (1982).
78. FUJIWARA, T., E. SUGISHITA, T. TAKEDA, Y. OGIHARA, M. SCHIMUZU, T. NOMURA, and Y. TOMITA: Further Studies on the Structure of Polysaccharides from the Bark of *Melia azadirachta*. Chem. Pharm. Bull., **32**, 1385 (1984).
79. FUJIWARA, T., T. TAKEDA, Y. OGIHARA, M. SCHIMUZU, T. NOMURA, and Y. TOMITA: Further Studies on the Structure of Polysaccharide from the Bark of *Melia azadirachta*. (III). Shoyakugaku Zasshi, **38**, 334 (1984).
80. GAITANDE, B. B., and V. K. SHETH: Pharmacological Studies of Sodium nimbidinate. Ind. J. Med. Sci., **12**, 156 (1958).
81. GAGNAIRE, D., and M. VINCENDEN: Easy Identification of Hydroxybearing Carbon Atoms in ^{13}C Nuclear Magnetic Resonance Spectroscopy – A new Method for Signal Assignment in Carbohydrates. J. Chem. Soc., Chem. Commun., 509 (1977).
82. GAIKWAD, B.R., T. MAYILVAGANAN, B.A. VYAS, and S.V. BHAT: Nimbocinol and 17-Epinimbocinol from the Nimbidin Fraction of Neem Oil. Phytochem., **29**, 3963 (1990).
83. GANDHI, M., R. LAL, A. SANKARANARAYANAN, C.K. BANERJEE, and P.L. SHARMA: Acute Toxicity Study of the Oil from *Azadirachta indica* Seed (Neem Oil). J. Ethnopharmacol., **23**, 39 (1988).
84. GARG, H.S., and D.S. BHAKUNI: An Isoprenylated Flavanone from Leaves of *Azadirachta indica*. Phytochem., **23**, 2115 (1984).
85. GARG, H.S., and D.S. BHAKUNI: Salannolide, a Meliacin from *Azadirachta indica*. Phytochem., **23**, 2383 (1984).
86. GARG, H.S., and D.S. BHAKUNI: 2′,3′-Dehydrosalannol, a Tetranortriterpenoid from *Azadirachta indica* (The Neem Tree). Phytochem., **24**, 866 (1985).
87. GLINSUKON, T., R. SOMJAREE, P. PIYACHATURAWAT, and Y. THEBTARANONTH: Acute Toxicity of Nimbolide and Nimbic Acid in Mice, Rats, and Hamsters. Toxicol. Letters, **30**, 159 (1986).
88. GOVINDACHARI, T.R.: Chemical and Biological Investigation on *Azadirachta indica* (The Neem Tree). Current Sci., **63**, 117 (1992).
89. GOVINDACHARI, T.R., G. GOPALAKRISHNAN, and G. SURESH: Isolation of Various Azadirachtins from Neem Oil by Preparative High Performance Liquid Chromatography. J. Liq. Chromatogr. And Related Technol., **19**, 1729 (1996).
90. GOVINDACHARI, T.R. and G. GOPALAKRISHNAN: 13,14-Desepoxyazadirachtin A – A New Tetranortriterpenoid from *Azadirachta indica*. Phytochem., **45**(2), 397 (1997).
91. GOVINDACHARI, T.R., G. GOPALAKRISHNAN, R. RAGHUNATHAN, and S.S. RAJAN: Crystallisation of Azadirachtin A. Current Sci., **66**, 295 (1994).
92. GOVINDACHARI, T.R., G. SANDHYA, and S.P. GANESHRAJ: Simple Method for the Isolation of Azadirachtin by Preparative High Performance Liquid Chromatography. J. Chromatogr., **513**, 389 (1990).

93. GOVINDACHARI, T.R., G. SANDHYA, and S.P. GANESHRAJ: Isolation of Novel Azadirachtins H and I by High Performance Liquid Chromatography. Chromatographia, **31**, 303 (1991).

94. GOVINDACHARI, T.R., G. SANDHYA, and S.P. GANESHRAJ: Azadirachtins H and I: Two New Tetranortriterpenoids from *Azadirachta indica*. J. Nat. Prod., **55**, 596 (1992).

95. GOVINDACHARI, T.R., G. SANDHYA, and S.P. GANESHRAJ: Structure of Azadirachtin K, a New Tetranortriterpenoid from *Azadirachta indica*. Indian J. Chem., **31B**, 295 (1992).

96. GRIFFITH, W.P., and S.V. LEY: TRAP – Tetra-*n*-propylammonium Perruthenate, a Mild and Convenient Oxidant for Alcohols. Aldrichim Acta, **23**, 13 (1990).

97. GRIFFITH, W.P., S.V. LEY, G.P. WHITCOMBE, and A.D. WHITE: Preparation and Use of Tetra-*n*-butylammonium Per-ruthenate (TBAP Reagent) and Tetra-n-propylammonium Per-ruthenate (TPAP Reagent) as New Catalytic Oxidants for Alcohols. J. Chem. Soc., Chem. Commun., 1625 (1987).

98. GROSSMAN, R.B., and S.V. LEY: Chemistry of Insect Antifeedants from *Azadirachta indica* (Part 16): Synthesis of Several Derivatives of Azadirachtin containing Fluorescent or Immunologenic Reporter Groups. Tetrahedron, **50** (29), 8871 (1994).

99. GROSSMAN, R.B., and S.V. LEY: Chemistry of Insect Antifeedants from *Azadirachta indica* (Part 17): Synthesis of Model Compounds of Azadirachtin. Unusual Effects of Remote Substituents on the Course of Oxidative Ring Contraction Reaction. Tetrahedron, **50** (39), 11553 (1994).

100. HARRIS, M., R. HENDERSON, R. MCCRINDLE, K.H. OVERTON, and D.W. TURNER: Tetranortriterpenoids – VIII. The Constitution and Stereochemistry of Nimbin. Tetrahedron, **24**, 1517 (1968).

101. HENDERSON, R., R. MCCRINDLE, A. MELERA, and K.H. OVERTON: Tetranortriterpenoids – IX. The Constitution and Stereochemistry of Salannin. Tetrahedron, **24**, 1525 (1968).

102. HAUNG, H.P., and E.D. MORGAN: Analysis of Azadirachtin by Super Critical Fluid Chromatography. J. Chromatogr., **519**, 137 (1990).

103. HIYAMA, T., K. NISHIDE, and K. K. KOBAYASHI: A New Synthesis of *N*-Benzoyl-*L*-Acosamine. Tetrahedron Letters, **25**, 569 (1984).

104. ISMAN, M.B., R. KOUL, A. LUCZYNSHI, and Z. KAMINSKI: Insecticidal and Antifeedant Bioactivities of Neem Oils and Their Relationship to Azadirachtin Content. J. Agric. Food Chem., **38**, 1406 (1990).

105. IWATA, M., and H. OHRAI: A Simple Regioselective Partial Hydrolysis of Di-*O*-isopropylidene Monosaccharides with Copper (II) ion. Bull. Chem. Soc. (Japan), **54**, 2837 (1981).

106. IWU, M.M., O. OBIDOA, and M. ANAZODO: Biochemical Mechanism of the Antimalarial Activity of *Azadirachta indica* Leaf Extract. Pharmac. Res. Commun., **18**, 81 (1986).

107. JACOBSON, M.: The Neem Tree – Natural Resistance par Excellence. Amer. Chem. Soc. Symp. Series No. 296, 221 (1986).

108. JONES, I.W., A.A. DENHOLM, S.V. LEY, H. LOVELL, A. WOOD, and R.E. SINDEN: Sexual Development of Malaria Parasites is inhibited *in vitro* by Neem Extract, Azadirachtin and its Semi-synthetic Analogues. FEMS Microbiology Letters, **120**, 267 (1994).

109. JONES, P.S., S.V. LEY, E.D. MORGAN, and D. SANTAFIANOS: The Chemistry of the Neem Tree. In: Focus on Phytochemical Pesticides Vol. 1 The Neem Tree (M. JACOBSON, ed.), p. 19. Boca Raton, FL (USA): CRC Press. 1989.

110. KABALEESWARAN, V., S.S. RAJAN, T.R. GOVINDACHARI, and G. GOPALAKRISHNAN: Crystal and Molecular Structure of Azadirachtin-A. Current Sci., **66** (5), 362 (1994).

111. KESSLER, H., C. GRIESINGER, J. ZARBOCK, and H.R. LOOSLI: Assignment of Carbonyl Carbons and Sequence Analysis in Peptides by Heteronuclear Shift Correlation via Small Coupling Constants with Broad band Decoupling in t_1 (COLOC). J. Magn. Reson., **57**, 331 (1984).

112. KHALID, S.A., H. DUDDECK, and M. GOZALEA-SIERRA: Isolation and Characterisation of an Antimalarial Agent of the Neem Tree *Azadirachta indica*. J. Nat. Prod., **52**, 922 (1989).

113. KIGODI, P.G.K., G. BLASKO, Y. THEBTARANONTH, J.M. PEZZUTO, and G.A. CORDELL: Spectroscopic and Biological Investigations of Nimbolide and 28-Deoxonimbolide from *Azadirachta indica*. J. Nat. Prod., **52**, 1246 (1989).

114. KLENK, A., M. BOKEL, and W. KRAUS: 3-Tigloylazadirachtol, an Insect Growth Regulating Constituent of *Azadirachta indica*. J. Chem. Soc., Chem. Commun., 523 (1986).

115. KOGA, Y., I. YOSHIDA, A. KIMURA, M. YOSHINO, F. YAMASHITA, and D. SINNIAH: Inhibition of Mitochondrial functions by Margosa Oil: Possible Implications in the Pathogenesis of Reye's Syndrome. Pediatric Res., **22**, 184 (1987).

116. KOLB, H.C., and S.V. LEY: Chemistry of Insect Antifeedants from *Azadirachta indica* (Part 10): Synthesis of a Highly Functionalised Decalin Fragment of Azadirachtin. Tetrahedron Letters, **32**, 6187 (1991).

117. KOOT, WIN-JAN, and S.V. LEY: Chemistry of Insect Antifeedants from *Azadirachta indica* (Part 18): Demethylation and Methylation of the C-8 Position of the Decalin Portion of Azadirachtin. Tetrahedron, **51**, 2077 (1995).

118. KOUL, O.: Feeding Deterrent Induced by Plant Limonoids in the Larvae of *Spodoptera litura* (F.) (Lepidoptera, Noctuidae). Z. Angew. Entomol., **95**, 166 (1983).

119. KOUL, O. and M.B. ISMAN: Toxicity of the Limonoid allelochemical cedrelone to noctuid larvae. Entomol. Exp. Appl., **64**, 281 (1992).

120. KOUL, O., M.B. ISMAN, and C.M. KETKAR: Properties and Uses of Neem, *Azadirachta indica*. Can. J. Bot., **68**, 1 (1990).

121. KRAUS, W.: Biologically Active Compounds from Meliaceae. Stud. Organ. Chem. **17**, 331 (1984).

122. KRAUS, W.: Constituents of Neem and Related Species. A revised Structure of Azadirachtin. Stud. Organ. Chem., **26**, 237 (1986).

123. KRAUS, W.: Constituents of Neem and Related Species. A Revised Structure of Azadirachtin. In: New Trends in Natural Product Chemistry (ATTA-UR-RAHMAN and P.W. LE QUESNE, eds.). Amsterdam, Netherlands: Elsvier Science Publishers B.V. 1986.

124. KRAUS, W.: Biologically Active Ingredients. In: The Neem Tree (H. SCHMUTTERER, ed.), p. 35. New York: VCH Publishers Inc. 1995.

125. KRAUS, W., S. BAUMANN, M. BOKEL, R. CRAMER, W. GRIMMINGER, R. HENDLEMEIER, E. KEIL, U. KELLER, A. KLENK, M.P.H. KLINGELE, and M. SCHWINGER: Insect Feeding and Development Controlling Constituents of Meliaceae. Proc. Ist Princess Chulabhorn Sci. Cong. Intern. Congr. on Nat. Products. Bangkok. II, 554 (1987).

126. KRAUS, W., S. BAUMANN, M. BOKEL, U. KELLER, A. KLENK, M. KLINGELE, H. POHNL, and M. SCHWINGER: Control of Insect Feeding and Development by Constituents of *Melia Azedarach* and *Azadirachta indica*. In: Natural Pesticides from the Neem Tree and Other Tropical Plants (H. SCHMUTTERER, and K.R.S. ASCHER, eds), p. 111. Eschborn, FRG: GTZ. 1987.

127. Kraus, W., and M. Bokel: Neue Tetranotriterpenoide aus *Melia azedarach* Linn. (Meliaceae). Chem. Ber., **114**, 267 (1981).
128. Kraus, W., M. Bokel, A. Bruhn, R. Cramer, I. Klaiber, A. Klenk, G. Nagi, H. Pohnl, H. Sadio, and B. Vogler: Structure Determination by NMR of Azadirachtin and Related Compounds from *Azadirachta indica* A. Juss. (Meliaceae). Tetrahedron, **43**, 2817 (1987).
129. Kraus, W., M. Bokel, R. Cramer, A. Klenk, and H.D. Pohnl: Constituents of Neem and Related Species. A Revised Structure of Azadirachtin. Third Int. Conf. on Chemistry and Biotechnology of Biologically Active Natural Products, Sofia, Bulgaria, p. 446 (1985).
130. Kraus, W., M. Bokel, A. Klenk, and H.D. Pohnl: The Structure of Azadirachtin and 22,23- Dihydro-23β-methoxyazadirachtin. Tetrahedron Letters, **26**, 6435 (1985).
131. Kraus, W., M. Bokel, M. Schwinger, B. Vogler, R. Soellner, D. Wendisch, R. Steffens, and U. Wachendorff: The Chemistry of Azadirachtin and Other Insecticidal Constituents of Meliaceae. In: Phytochemistry and Agriculture (T. In Van Beek, H. Breteler, eds.), p. 18–39. Oxford University Press. 1993.
132. Kraus, W., M. Bokel, R. Soellner, B. Vogler, D. Wendisch, and Y. Zhou-Halwart: Deoxatriquinane Derivatives as Intermediates in the Biosynthesis of Azadirachtin. 9th Annual Meeting of the Int. Soc. of Chemical Ecology (Kyoto, Japan), p. 55, 1992; 18th IUPAC Annual Symp. on the Chemistry of Natural Products, Strasbourg, France, p. 351 (1992).
133. Kraus, W., and R. Cramer: 17-Epiazadiradion und 17-Hydroxyazadiradion Zwei Neue Inhaltsstoffe aus *Azadirachta indica* A. Juss. Tetrahedron Letters, 2395 (1978).
134. Kraus, W., and R. Cramer: Neue Tetranortriterpenoide mit Insektenfrashemmender Wirkung aus Neem Oil. Liebigs Ann. Chem., 181 (1981).
135. Kraus, W., and R. Cramer: Pentanortriterpenoide aus *Azadirachta indica* A. Juss (Meliaceae). Chem. Ber., **114**, 2375 (1981).
136. Kraus, W., R. Cramer, and G. Sawitzki: Tetranortriterpenoids from the Seeds of *Azadirachta indica*. Phytochem., **20**, 117 (1981).
137. Kraus, W., H. Gutzeit, and M. Bokel: 1,3-Diacetyl-11-19-deoxa-11-oxomeliacarpin, a Possible Precursor of Azadirachtin, from *Azadirachta indica* A. Juss (Meliaceae). Tetrahedron Letters, **30**, 1797 (1989).
138. Kraus, W., A. Klenk, M. Bokel, and B. Vogler: Tetranortriterpenoid-Lactams with Insect Feeding Deterrent Effect from *Azadirachta indica* A. Juss (Meliaceae). Liebigs Ann. Chem., 337 (1987).
139. Kubo, I., A. Matsumoto, and T. Matsumoto: New Insect Ecdysis Inhibitory Limonoid Deacetyl Azadirachtinol Isolated from *Azadirachta indica* (Meliaceae) Oil. Tetrahedron, **42**, 489 (1986).
140. Kubo, I., T. Matsumoto, A. Matsumoto, and J.N. Shoolery: Structure of Deacetylazadirachtinol, Application of 2D ^{1}H-^{1}H and ^{1}H-^{13}C Shift Correlation Spectroscopy. Tetrahedron Letters, **25**, 4729 (1984).
141. Kumar, Ch. S.S.R., M. Srinivas, and S. Yukkundi: Limonoids from the Seeds of *Azadirachta indica*. Phytochem., **43** (2), 451 (1996).
142. Lakshmi, S.U., and T.N. Pattabiraman: Studies on Plant Gums: Part I. Identification of Nitrogenous Compounds in Neem (*Azadirachta indica*) Gum and Isolation of D-Glucosamin. Indian J. Biochem., **4**, 183 (1967).
143. Lavie, D., and M.K. Jain: Tetranortriterpenoids from *Melia azadirachta* L. J. Chem. Soc., Chem. Commun., **D**, 278 (1967).
144. Lavie, D., M.K. Jain, and S.R. Shpan-Gabrielith: A Locust Phagorepellent from two *Melia* species. Chem. Commun., 910 (1967).

145. LAVIE, D., M.K. JAIN, and I. KIRSON: Terpenoids. Part VI. The Complete Structure of Melianone. J. Chem. Soc (C), 1347 (1967).
146. LAVIE, D., and E.C. LAVY: Studies on Epoxides IV. Rearrangements in Triterpenoids. Tetrahedron Letters, 2097 (1968).
147. LAVIE, D., and E.C. LAVY: A Compound Linking Melianes with Meliacins. Tetrahedron Letters, 3525 (1969).
148. LAVIE, D., and E.C. LAVY: Oxidative Reactions of Biogenetic Interest. Tetrahedron Letters, 1315 (1970).
149. LAVIE, D., and E.C. LAVY: Meliane-Meliacin Relationship. Tetrahedron, **27**, 3941 (1971).
150. LAVIE, D., E.C. LAVY, and M.K. JAIN: Limonoids of Biogenetic Interest from *Melia Azadirachta* L. Tetrahedron, **27**, 3927 (1971).
151. LEE, S.M., and J.A. KLOCKE: Combined Florisil, Droplet Counter Current and High Performance Liquid Chromatographies for Preparative Isolation and Purification of Azadirachtin from Neem (*Azadirachta indica*) Seeds. J. Liq. Chromatogr., **10**, 1151 (1987).
152. LEE, S.M., J.A. KLOCKE, M.A. BARNBY, R.B. YAMASAKI, and M.F. BALADRIN: Insecticidal Constituents of *Azadirachta indica* and *Melia Azadirach* (Meliaceae). In: Naturally Occurring Pest Bioregulators (P.A. HEDIN, ed.), ACS Symposium Series No. **449**, p. 293. Washington, DC: American Chemical Society, 1991.
153. LEE, S.M., J.L. OLSEN, M.P. SCHWEIZER, and J.A. KLOCKE: 7-Deacetyl-17β-hydroxyazadiradione, a New Limonoid Insect Growth Inhibitor from *Azadirachta indica*. Phytochem., **27**, 2773 (1988).
154. LEY, S.V.: Synthesis of Insect Antifeedants. In: Pesticide Science ans Biotechnology (R. GREENHALGH and T.R. ROBERTS, eds.). 1987.
155. LEY, S.V.: Insect Antifeedants. In: Pesticide Chemistry (H. FREHSE, ed.) p. 97. New York: VCH. 1990. Proceedings of the Seventh International Congress of Pesticide Chemistry, Hamburg, 1990.
156. LEY, S.V., J.C. ANDERSON, W.M. BLANEY, P.S. JONES, Z. LIDERT, E.D. MORGAN, N.G. ROBINSON, D. SANTAFIANOS, M.S.J. SIMMONDS, and P.L. TOOGOOD: Insect Antifeedants from *Azadirachta indica* (Part 5): Chemical Modification and Structure-Activity Relationships of Azadirachtin and Some Related Limonoids. Tetrahedron, **45**, 5175 (1989).
157. LEY, S.V., J.C. ANDERSON, W.M. BLANEY, Z. LIDERT, E.D. MORGAN, N.G. ROBINSON, and M.S.J. SIMMONDS: Chemistry of Insect Antifeedants from *Azadirachta indica* (Part 3): Reaction on the C-22,23 Enolether Double Bond of Azadirachtin and Conversion to 22,23-Dihydro-23β-methoxyazadirachtin. Tetrahedron Letters, **29**, 5433 (1988).
158. LEY, S.V., J.C. ANDERSON, W.M. BLANEY, E.D. MORGAN, R.N. SHEPPARD, M.S.J. SIMMONDS, A.M.Z. SLAWIN, S.C. SMITH, D.J. WILLIAMS, and A. WOOD: Chemistry of Insect Antifeedants from *Azadirachta indica* (Part 11): Characterisation and Structure-Activity Relationships of Some Novel Rearranged Azadirachtins. Tetrahedron, **47**, 9231 (1991).
159. LEY, S.V., A.A. DENHOLM, and A. WOOD: Chemistry of Azadirachtin. Nat. Prod. Rep., 109 (1993).
160. LEY, S.V., K. DOHERTY, G. MASSIOT, and J.-M. NUZILLARD: "Connectivist" Approach to Organic Structure Determination Lsd-Program Assisted Nmr Analysis of the Insect Antifeedant Azadirachtin. Tetrahedron, **50**, 12267 (1994).
161. LEY, S. V., P. J. LOVELL, A.M.Z. SLAVIN, S.C. SMITH, D.J. WILLIAMS, and A. WOOD: Chemistry of Insect Antifeedants from *Azadirachta indica* (Part 15). Tetrahedron, **49**, 1675 (1993).

162. LEY, S.V., P.J. LOVELL, S.C. SMITH, and A. WOOD: Chemistry of Insect Antifeedants from *Azadirachta indica* (Part 9): Oxidative Reactions of Azadirachtin Derivatives leading to C(8)-C(14) Bond Cleavage. Tetrahedron Letters, **32**, 6183 (1991).

163. LEY, S.V., H. LOVELL, and D.J. WILLIAMS: Chemistry of Insect Antifeedants from *Azadirachta indica*, Part 14: Absolute Configuration of Azadirachtin. J. Chem. Soc., Chem. Commun., 1304 (1992).

164. LEY, S.V., D. SANTAFIANOS, W.M. BLANEY, and M.S.J. SIMMONDS: Synthesis of a Hydroxydihydrofuranacetal Related to Azadirachtin – A Potent Insect Antifeedant. Tetrahedron Letters, **28**, 221 (1987).

165. LEY, S.V., A.A. SOMOVILLA, H.B. BROUGHTON, D. CRAIG, A.M.Z. SLAWIN, P.L. TOOGOOD, and D.J. WILLIAMS: Chemistry of Insect Antifeedants from *Azadirachta indica* (Part 4): Synthesis towards the Limonoid Azadirachtin: Preparation of a Functionalised Decalin Fragment. Tetrahedron, **45**, 2143 (1989).

166. LYONS, C.W., and D.R. TAYLOR: Terpenoids. Part VI. The Complete Structure of Melianone. J. Chem. Soc., Chem. Commun., 517 (1975).

167. LUSCOMBE, D.K., and S.A. TAHA: Pharmacological Studies on the Leaves of *Azadirachta indica*. J. Pharmacy and Pharmacology, **26**, Suppl., 111 (1974).

168. MADHUSUDANAN, K.P., R. CHATURVEDI, H.S. GARG, and D.S. BHAKUNI: Negative Ion Mass Spectra of Tetranortriterpenoids Isolated from Neem (*Azadirachta indica* A. JUSS). Indian J. Chem., **23B**, 1082 (1984).

169. MAJUMDAR, P.L., D.C. MAITI, W. KRAUS, and M. BOKEL: Nimbidiol, a Modified Diterpenoid of the Root-bark of *Azadirachta indica*. Phytochem., **26**, 3021 (1987).

170. MANCUSO, A.J., and D. SWERN: Activated Dimethyl Sulfoxide – Useful Reagents for Synthesis. Synthesis, 165 (1981).

171. MEINWALD, J., and E. FRAUENGLASS: A Baeyer – Villiger Oxidation of Bicyclic Ketones. J. Amer. Chem. Soc., **82**, 5235 (1960).

172. MITRA, C.R.: On the Constitution of Nimbin. J. Sci. Indust. Res., **16B**, 477 (1957).

173. MITRA, C.R., H.S. GARG, and G.N. PANDEY: Constituents of *Melia indica* II. Nimbidic Acid and Nimbidinin. Tetrahedron Letters, 2761 (1970).

174. MITRA, C.R., H.S. GARG, and G.N. PANDEY: Identification of Nimbidic Acid and Nimbidinin from *Azadirachta indica*. Phytochem., **10**, 857 (1971).

175. MORGAN, E.D. In: Natural Pesticides from the Neem Tree (*Azadirachta indica* A. JUSS.) (H. SCHMUTTERER, K.R.S. ASCHER, and H. REMBOLD, eds.), Proceedings of the First International Neem Conference, p. 43. Eschborn, Germany: German Agency for Technical Cooperation. 1982.

176. MORGAN, E.D., and M.D. THORNTON: Azadirachtin in the Fruit of *Melia azadirach*. Phytochemistry, **12**, 391 (1973).

177. MUBARAK, A.M., and C.P. KULATILLEKE: Sulphur Constituents of Neem Seed Volatiles: A Revision. Phytochem., **29**, 3351 (1990).

178. MORI, K., and H. WATANABE: IUPAC 7th International Congress of Pesticide Chemistry, Hamburg, Book of Abstracts, Vol. I, p. 1251 (1990).

179. MUKHERJEE, S., and H.C. SRIVASTAVA: Structure of Neem Gum. J. Amer. Chem. Soc., **77**, 422 (1955).

180. MURTY, K.S., D.N. RAO, D.K. RAO, and L.B.G. MURTY: A preliminary Study on Hypoglycaemic and Antihyperglycaemic Effects of *Azadirachta indica*. Ind. J. Pharmacol., **10**, 247 (1978).

181. NAKANISHI, K.: Structure of Insect Antifeedant Azadirachtin. Recent Adv. Phytochem., **9**, 283 (1975).

182. Nanduri, Srinivas, and K. P. Banstola: Neeflone, a New Tetranortriterpenoid from the Flowers of *Azadirachta indica* A. Juss. (Meliaceae). Indian J. Chem., Sect. B, **34B**(11), 1019 (1995).

183. Narayanan, C.R., and K.N. Iyer: Isolation and Characterisation of Deacetylnimbin. Indian J. Chem., **5**, 460 (1967).

184. Narayanan, C.R., R.V. Pachapurkar, S.K. Pradhan, and V.R. Shah: Structure of Nimbin. Chem. and Ind., 22nd Feb, 322 (1964).

185. Narayanan, C.R., R.V. Pachapurkar, S.K. Pradhan, V.R. Shah, and N.S. Narasimhan: Structure of Nimbin. Indian J. Chem., **2**, 108 (1964).

186. Narayanan, C.R., R.V. Pachapurkar, and B.M. Sawant: Nimbinin – A New Tetranortriterpenoid. Tetrahedron Letters, 3563 (1967).

187. Narayanan, C.R., R.V. Pachapurkar, B.M. Sawant, and M.S. Wadia: Vepinin, A New Constituent of Neem Oil. Indian J. Chem., **7**, 187 (1969).

188. Nath, B.: Chemical Examination of the Heartwood of *Melia azadirachta*. J. Sci. Indust. Res, **14B**, 634 (1955).

189. National Research Council: Neem – A Tree for Solving Global Problems. Washington, DC: National Academic Press. 1992.

190. Nayak, B.R., and T.N. Pattabiraman: Studies on Plant Gums. Part III: Isolation and Characterisation of a Glycopeptide from Neem (*Azadirachta indica*) Gum after Pronase Digestion. Indian J. Biochem. Biophys., **15**, 449 (1978).

191. Obaseki, O., and H.A. Jegede-Fagunsin: The Antimalarial Activity of *Azadirachta indica*. Fitoterapia, **57**, 247 (1986).

192. Obih, P.O., and J.M. Makinde: Effect of *Azadirachta indica* on *Plasmodium berghei* in Mice. African J. Medicine and Medical Sciences, **14**, 51 (1985).

193. Ochi, M., T. Kotsuki, T. Tada, and T. Tokoroyama: Limonoids from *Melia azedarach* Linn. Var. *Japonica makino* III. The Structure of Ochinal and Ochinin acetate. Chem. Letters. 331 (1978).

194. Okpako, D.T.: Prostaglandin Synthatase Inhibitory Effect of *Azadirachta indica*. J. West African Sci. Assoc., **22**, 45 (1977).

195. Okpanyi, S.N., and G.C. Ezeukwu: Anti-inflammatory and Antipyretic Activities of *Azadirachta indica*. Planta Medica, **41**, 34 (1981).

196. Pachapurkar, R.V., and P.M. Kornule: Tetranortriterpenoids from the Leaves of *Azadirachta indica*. Acta Ciencia Indica, **9c**, 55 (1983).

197. Pachapurkar, R.V., P.M. Kornule, and C.R. Narayanan: A New Hexacyclic Tetranortriterpenoid. Chem. Letters, **4**, 357 (1974).

198. Pant, N., H.S. Garg, K.P. Madhusudanan, and D.S. Bhakuni: Sulfurous Compounds from *Azadirachta indica* Leaves. Fitoterapia, **57**, 302 (1986).

199. Pfileger, D., B. Muckensturm, P.C. Robert, M.T. Simonis, and J.C. Klenlen: Synthesis of the Furo-pyran Moiety of Azadirachtin. Tetrahedron Letters, **28**, 1519 (1987).

200. Pillai, N.R., D. Sugantla, C. Seshadri, and G. Santhakumari: Anti-gastric Ulcer Activity of Nimbidin. Indian J. Med. Res. **68**, 1969 (1978).

201. Pillai, N.R., and G. Santhakumari: Anti-arthritic and Anti-inflammatory Actions of Nimbidin. Planta Medica, **43**, 59 (1981).

202. Pillai, N.R., and G. Santhakumari: Hypoglycaemic Activity of *Melia azadirachta* Linn. (Neem). Indian J. Med. Res., **74**, 931 (1981).

203. Pillai, N.R., and G. Santhakumari: Effect of Nimbidin on Acute and Chronic Gastro-duodenal Ulcer Models in Experimental Animals. Planta Medica, **50**, 143 (1984).

204. PILLAI, N.R., and G. SANTHAKUMARI: Toxicity Studies on Nimbidin, a Potential Antiulcer Drug. Planta Medica, **50**, 146 (1984).

205. PODDER, G., and S.B. MOHATO: Azadirachtanin, A New Limonoid from the Leaves of *Azadirachta indica*. Heterocycles, **23**, 2321 (1985).

206. POLONSKY, J., Z. BASKEVITCH, H.E. GOTTLIEB, E.W. HAGAMAN, and E. WENKERT: Carbon-13 Nuclear Magnetic Resonance Spectral Analysis of Quassinoid Bitter Principles. J. Organ. Chem., **40**, 2499 (1975).

207. PRAKASH, A.O., R.K. TEWARI, and R. MATHUR: Non-hormonal Post-coital Contraceptive Action of Neem Oil in Rats. J. Pharmacol., **23**, 53 (1988).

208. RAGASA, C.Y., Z.D. NACPIL, G.M. NATIVIDAD, M. TADA, J.C. COLL, and J.A. RIDEOUT: Tetranortriterpenoids from *Azadirachta indica*. Phytochem., **46**(3), 555 (1997).

209. RAJAB, M.S., M.D. BENTLEY, and R.C. FORT Jr.: Biomimetic Formation of a Nimbin Class Limonoid. J. Nat. Prod., **51**, 1291 (1988).

210. RAMESH, K., and M.A. PADHYA: Isolation of Nimbin from *Azadirachta indica* Leaves and Its Callus Culture. Indian Drugs, **25**, 526 (1988).

211. RAMJI, N., K. VENKATAKRISHNAN, and K.M. MADYASTHA: 11-Epi-azadirachtin H from *Azadirachta indica*. Phytochem., **42**(2), 561 (1996).

212. RANI, K., and A. AKHILA, Biosynthetic Relationship between Nimocinol and Nimocinolide in *Azadirachta indica*. Nat. Prod. Letters, **4**, 179 (1994)

213. RAO, B.S., NAZMA, and J.M. RAO: Anti-fungal Activity of Gedunin. Current Sci., **46**, 714 (1977).

214. RAO, K.N., and B.S. PARMAR: A Compendium of Chemical Constituents of Neem. Neem Newsletter, **1**, 39 (1984).

215. RAO, D.V.K., K. SINGH, P. CHOPRA, P.C. CHABRA, and G. RAMANUJALU: *In vitro* Antibacterial Activity of Neem Oil. Indian J. Med. Res., **84**, 314 (1986).

216. RAO, A.R., S. SUKUMAR, T.V. PARAMASIVAM, S. KAMALAKSHI, A.R. PARASHURAMAN, and M. SHANTA: Study of Antiviral Activity of Tender Leaves of Margosa Tree (*Melia azadirachta*) on Vaccinia and Variola Virus – A Preliminary Report. Indian J. Med. Res., **57**, 495 (1969).

217. REMBOLD, H.: Isomeric Azadirachtins and their Mode of Action. In: Focus on Phytochemical Pesticides. Vol 1, The Neem Tree (M. JACOBSON, ed.), p. 47. Boca Raton, FL: CRC Press. 1988.

218. REMBOLD, H.: The Azadirachtins – Their Potential for Insect Control. In: Economic and Medicinal Plant Research, Vol.3, p.57 (1989).

219. REMBOLD, H.: Azadirachtins: Their Structure and Mode of Action. In: Insecticides of Plant Origin (J.T. ARNASON, B.J.R. PHILOGENE and P. MORAND, eds.), p. 150. ACS Sym. Ser. 387. Washington, DC: American Chemical Society. 1989.

220. REMBOLD, H.: Secondary Plant Products in Insect Growth Control, with Special Reference to Azadirachtins. Adv. Invertebr. Reprod. **3**, 481 (1984).

221. REMBOLD, H.: Biochemistry of Neem. World Neem Conf., Bangalore, India, p. 53, 1993.

222. REMBOLD, H., H. FORSTER, and C. CZOPPELT: Structure and Biological Activity of Azadirachtins A and B. In: Natural Pesticides from the Neem Tree and Other Tropical Plants (H. SCHMUTTERER and K.R.S. ASCHER, eds), p. 149. 1987; Proc. 3rd Neem Conf., Nairobi, Kenya, 1986.

223. REMBOLD, H., H. FORSTER, and J. SONNENBICHLER: Structure of Azadirachtin B. Z. Naturforsch., **42C**, 4 (1987).

224. RIAR, S.S., C. DEVAKUMAR, G. ILAVAZHAGAN, J. BARDHAN, A.K. KAIN, P. THOMAS, R. SINGH, and B. SINGH: Volatile Fraction of Neem Oil as a Spermicide. Contraception, **42**, 479 (1990).

225. ROCHANAKIJ, S., Y. THEBTARANONTH, and C. TENJAI: Nimbolide, a Constituent of *Azadirachta indica* inhibits *Plasmodium falciparum* in Culture. Southeast Asian J. Trop. Med. and Public Health, **16**, 66 (1985).
226. ROJANAPO, W., S. SUWANNO, R. SOMJAREE, T. GLINSUKON, and Y. THEBTARANONTH: Mutagenic and Antibacterial Activity Testing in Nimbolide and Nimbic Acid. J. Sci. Soc. Thailand, **11**, 177 (1985).
227. ROJATKAR, S.R., V.S. BHAT, M.M. KULKARNI, V.S. JOSHI, and B.A. NAGASAMPAGI: Tetranortriterpenoids from *Azadirachta indica*. Phytochem., **28**, 203 (1989).
228. ROJATKAR, S.R., and B.A. NAGASAMPAGI: 1-Tigloyl-3-acetyl-11hydroxy-4β-methylmeliacarpin from *Azadirachta indica*. Phytochemistry, **32**, 213 (1993).
229. ROJATKAR, S.R., and B.A. NAGASAMPAGI: 11α-Hydroxy-12-norazadirachtin from *Azadirachta indica*. Nat. Prod. Letters, **5(1)**, 69 (1994).
230. ROJATKAR, S.R., and B.A. NAGASAMPAGI: 3α-Acetoxy-1α-hydroxyazadirachtol, a New Constituent from *Azadirachta indica*. Indian J. Chem., Sect. B, **34B**(11), 1016 (1995).
231. ROJATKAR, S.R., D.D. SAWAIKAR, and B.A. NAGASAMPAGI: New Tetra and Pentanortriterpenoids from *Azadirachta indica* A. JUSS. World Neem Conference, India, 203 (1993).
232. RUKMINI, C.: Chemical and Nutritional Evaluation of Neem Oil. Food Chem., **26**, 119 (1987).
233. SANKARAM, A.V.B., M. MARTHANDAMURTHY, K. BHASKARIAH, M. SUBRAMANYAM, N. SULTANA, H.C. SHARMA, and K. LEISCHNER: 16th International Symposium on the Chemistry of Natural Products, IUPAC, Japan, Kyoto, 1988, Abs. No. 1539, p. 41.
234. SANKARAM, A.V.B., M. MARTHANDAMURTHY, K. BHASKARIAH, M. SUBRAMANYAM, N. SULTANA, H.C. SHARMA, K. LEUSCHNER, G. RAMAPRASAD, S. SITARAMAIAH, C. RUKMINI, and P.U. RAO: Chemistry, Biological Activity and Utilisation Aspects of Some Promising Neem Extractives. In: Natural Pesticides from the Neem Tree and Other Tropical Plants (H. Schmutterer and K.R.S. Ascher eds.), p. 127–148. Eschborn, FRG: GTZ. 1987.
235. SANTHAKUMARI, G., and J. STEPHEN: Antimiototic Effect of Nimbidin – A First Report. Experientia, **37**, 91(1981).
236. SANYAL, M. and P.C. DUTTA: Nimbin Biosynthesis and the Age of Cultured Callus from Neem Bark. Indian Drugs, **19**, 61 (1981).
237. SANYAL, M., S. TIKADAR, and P.C. DUTTA: Nimbin and β-Sitosterol in Cotyledon and Cultured Tissues of *Azadirachta indica*. Indian Drugs, **20**, 479 (1983).
238. SARKAR, M.S., and P.C. DATTA: Age Factor in Biosynthesis of Nimbin and β-Sitosterol in the Bark and Callus of *Azadirachta indica*. Indian Drugs, **24**(1), 62 (1986).
239. SARKAR, M.S., A. MUKHERJI, and P.C. DATTA: Glycine on *in vitro* Biosynthesis of Nimbin and β-Sitosterol in Tissues of *Azadirachta indica*. Current Sci., **57**, 40 (1988).
240. SATO, T., J. OTERA, and H. NOZAKI: Activation and Synthetic Applications of Thiostannanes. Thioalkoxylation of Acetals. Tetrahedron, **45**, 1209 (1989).
241. SATO, Y., M. SODEOKA and M. SHIBASAKI: Catalytic Asymmetric C–C Bond Formation: Asymmetric Synthesis of *cis*-Decalin Derivatives by Palladium-catalysed Cyclisation of Prochiral Alkenyl Iodides. J. Organ. Chem., **54**, 4738 (1989).
242. SATO, Y., M. SODEOKA and M. SHIBASAKI: On the Role of Silver Salts in Asymmetric Heck-type Reaction – A Greatly Improved Catalytic Asymmetric Synthesis of *cis*-Decalin Derivatives. Chem. Letters, 1953 (1990).
243. SATO, Y., S. WATANABE, and H. SHIBASAKI: Further Studies on a Catalytic Asymmetric Synthesis of Decalin Derivatives. Tetrahedron Letters, **33**, 2589 (1992).

244. SCHNEIDER, B.H., and K. ERMEL: Quantitative Determination of Azadirachtin from Neem Seeds using High Performance Liquid Chromatography. In: Natural Pesticides from the Neem Tree and Other Tropical Plants (H. SCHMUTTERER and K.R.S. ASCHER eds.), p. 161–170. Eschborn, FRG: GTZ. 1987; Proc. 3rd Int. Neem Conf., Nairobi, Kenya, 1986.

245. SCHROEDER, D.R., and K. NAKANISHI: A Simplified Isolation Procedure for Azadirachtin. J. Nat. Prod., **50**, 241 (1987).

246. SEEBACH, D.: Angew Chem., Int. Ed. Engl., **18**, 239 (1979).

247. SENGUPTA, P., S.N. CHAUDHURI, and H.N. KHASTIGIR: Terpenoids and Related Compounds I. Constituents of the Trunk Bark of *Melia azadirachta* Linn. and the Structure of the Ketophenol, Nimbiol. Tetrahedron, **10**, 45 (1960).

248. SENGUPTA, P., S.K. SENGUPTA, and H.N. KASTIGIR: Terpenoids and Related Compounds II. Investigations on Structure of Nimbin. Tetrahedron, **11**, 67 (1960).

249. SHARPLESS, K.B., M.A. UMBREIT, M.T. NIEH, and T.C. FLOOD: Lower Valent Tungsten Halides – A New Class of Reagents for Deoxygenation of Organic Molecules. J Amer. Chem. Soc., **94**, 6538 (1972).

250. SHUKLA, R., S. SINGH, C.R. BHANDARI: Preliminary Clinical Trials on Antidiabetic Actions of *Azadirachta indica*. Medicine and Surgery **13**, 11 (1973).

251. SIDDIQUI, S.: A Note on the Isolation of Three New Bitter Principles from the Nim Oil. Current Sci., **11**, 278 (1942).

252. SIDDIQUI, S., I. ARA, S. FAIZI, T. MAHMOOD, and B.S. SIDDIQUI: Phenolic Tricyclic Diterpenoids from the Bark of *AAzadirachta indica*. Phytochem., **27**, 3903 (1988).

253. SIDDIQUI, S., S. FAIZI, T. MAHMOOD, and B.S. SIDDIQUI: Isolation of a New Tetranortriterpenoid from *Azadirachta indica* A. JUSS (Meliaceae). Heterocycles, **24**, 1319 (1986).

254. SIDDIQUI, S., S. FAIZI, T. MAHMOOD, and B.S. SIDDIQUI: Two New Insect Growth Regulator Meliacins from *Azadirachta indica* A. JUSS (Meliaceae). J. Chem. Soc. Perkin Trans. **1**, 1021 (1986).

255. SIDDIQUI, S., S. FAIZI, T. MAHMOOD, and B.S. SIDDIQUI: Margosinolide and Isomargosinolide, Two New Tetranortriterpenoids from *Azadirachta indica* A. JUSS (Meliaceae). Tetrahedron, **42**, 4849 (1986).

256. SIDDIQUI, S., S. FAIZI, and B.S. SIDDIQUI: Studies on the Chemical Constituents of *Azadirachta indica* A. JUSS (Meliaceae). 1. Isolation and Structure of a New Tetranortriterpenoid, Nimolicinol. Heterocycles, **22**, 295 (1984).

257. SIDDIQUI, S., S. FAIZI, and B.S. SIDDIQUI: Studies on the Chemical Constituents of *Azadirachta indica* A. JUSS (Meliaceae). VII. Z. Naturforsch., **42B**, 922 (1986).

258. SIDDIQUI, S., S. FAIZI, B.S. SIDDIQUI, and GHIASUDDIN: Constituents of *Azadiachta indica*: Isolation and Structure Elucidation of a New Antibacterial Tetranortriterpenoid, Mahmoodin, and a New Protolimonoid, Naheedin. J. Nat. Prod., **55**, 303 (1992).

259. SIDDIQUI, S., S. FUCHS, J. LUCKE, and W. VOELTER: Structure of a New Natural Product from *Melia azadirachta* Linn.: 17-Hydroxyazadiradione. Tetrahedron Letters, 611 (1978).

260. SIDDIQUI, B.S., GHIASUDDIN, S. FAIZI, and S. SIDDIQUI: Triterpenoids from the Fresh Fruit Coats of *Azadirachta indica*. Phytochemistry, **31**, 4275 (1992).

261. SIDDIQUI, S., GHIASUDDIN, B.S. SIDDIQUI, and S. FAIZI: Tetranortriterpenoids and Steroidal Glycosides from the Seeds of *Azadirachta indica* A. JUSS. Pak. J. Sci. Indust. Res., **32**, 435 (1989).

262. SIDDIQUI, S., GHIASUDDIN, B.S. SIDDIQUI, and S. FAIZI: Triterpenoids from Kernel of *Azadirachta indica*. Proc. Pak. Acad. Sci., **27**, 333 (1990).

263. SIDDIQUI, S., T. MAHMOOD, B.S. SIDDIQUI, and S. FAIZI: Studies on the Norterpenoidal Constituents of *Azadirachta indica*. Pak. J. Sci. Indust. Res., **28**, 1 (1985).
264. SIDDIQUI, S., T. MAHMOOD, B.S. SIDDIQUI, and S. FAIZI: Isolation of a Triterpenoid from *Azadirachta indica*. Phytochem., **25**, 2183 (1986).
265. SIDDIQUI, S., T. MAHMOOD, B.S. SIDDIQUI, and S. FAIZI: Two New Tetranortriterpenoids from *Azadirachta indica*. J. Nat. Prod., **49**, 1068 (1986).
266. SIDDIQUI, S., T. MAHMOOD, S. FAIZI, and B.S. SIDDIQUI: Studies in the Chemical Constituents of *Azadirachta indica* A. JUSS (Meliaceae) X. Isolation and Structure Elucidation of Isonimolicinolide, the First 17-Acetoxy tetranortriterpenoid and nimolicinoic acid, the First Hexanortriterpenoid with an Apoeuphane (Apotirucallane) Skeleton. J. Chem. Soc. Perkin Trans. **1**, 1429 (1987).
267. SIDDIQUI, S., T. MAHMOOD, B.S. SIDDIQUI, and S. FAIZI: Isonimolide and Isonimbolide, Two New Tetranortriterpenoids from the Twigs of *Azadirachta indica* A. JUSS (Meliaceae). Heterocycles, **26**(7), 1827 (1987).
268. SIDDIQUI, S., T. MAHMOOD, B.S. SIDDIQUI, and S. FAIZI: Non-terpenoidal Constituents from *Azadirachta indica*. Planta Med., **54**, 457 (1988).
269. SIDDIQUI, S. and C.R. MITRA: Utilisation of Nim Oil and its Bitter Constituents (Nimbidin series) in the Pharmaceutical Industry. J. Sci. Indust. Res. **4**, 5 (1945).
270. SIDDIQUI, S., B.S. SIDDIQUI, S. FAIZI, and T. MAHMOOD: Isolation of a Tetranortriterpenoid from *Azadirachta indica*. Phytochem., **23**, 2899 (1984).
271. SIDDIQUI, S., B.S. SIDDIQUI, and S. FAIZI: Studies in the Chemical Constituents of *Azadirachta indica* 2. Isolation and Structure Elucidation of the New Terpenoid, Azadirachtol. Planta Medica, **51**, 478 (1985).
272. SIDDIQUI, S., B.S. SIDDIQUI, S. FAIZI, and T. MAHMOOD: Isoazadirolide, a New Tetranortritepenoid from *Azadirachta indica* A. JUSS (Meliaceae). Heterocycles, **24**, 3163 (1986).
273. SIDDIQUI, S., B.S. SIDDIQUI, S. FAIZI and T. MAHMOOD: Studies on the Chemical Constituents of *Azadirachta indica* A. JUSS (Meliaceae). VI. J. Chem. Soc. Pakistan, **8**, 341 (1986).
274. SIDDIQUI, S., B.S. SIDDIQUI, S. FAIZI, and T. MAHMOOD: Tetracyclic Triterpenoids and their Derivatives from *Azadirachta indica*. J. Nat. Prod., **51**, 30 (1988).
275. SIDDIQUI, S., B.S. SIDDIQUI, T. MAHMOOD, and S. FAIZI: Tetranortriterpenoids from *Azadirachta indica* A. JUSS (Meliaceae). Heterocycles, **29**, 87 (1989).
276. SIDDIQUI, S., B.S. SIDDIQUI, GHIASUDDIN, and S. FAIZI: New Meliacin Analogues from Epoxyazadiradione. Pak. J. Sci. Indust. Res., **33**, 359 (1990).
277. SIDDIQUI, S., B.S. SIDDIQUI, GHIASUDDIN, and S. FAIZI: Terpenoids from Fruit Coating of *Azadirachta indica*. Phytochem., **30**, 1615 (1991).
278. SIDDIQUI, S., B.S. SIDDIQUI, GHIASUDDIN, and S. FAIZI: Tetracyclic Triterpenoids of the Fruit coats of *Azadirachta indica*. J. Nat. Prod., **54**, 408 (1991).
279. SIDDIQUI, S., T.N. WAHEED, S. FUCHS, J. LUCKE, and W. VOELTER: The Structure of a New Compound Isolated from the Fruit Pulp of *Melia azadirachta* Linn. Z. Naturforsch., **30b**, 961 (1975).
280. SIMMONDS, M.S.J., W.M. BLANEY, S.V. LEY, J.C. ANDERSON, and P.L. TOOGOOD: Azadirachtin Structural Requirements for Reducing Growth and Increasing Mortality in Lepidopterous Larvae. Entomol. Exp. Appl., **55**, 169 (1990).
281. SINGH, P.P., A.Y. JUNNARKAR, G.S. REDDI, K.V. SINGH: *Azadirachta indica* – Neuropsychopharmacological and Antimicrobial Studies. Fitoterapia, **58**, 235 (1987).
282. SINHA, K.C., S.S. RIAR, J. BARDHAN, P. THOMAS, A.K. JAIN, and R.K. JAIN: Antiimplantation Effect of Neem Oil. Indian J. Med. Res., **80**, 708 (1984).

283. SINHA, K.C., S.S. RIAR, R.S. TIWARY, A.K. DHAWAN, J. BARDHAN, P. THOMAS, A.K. KAIN, and R.K. JAIN: Neem Oil as a Veginal Contraceptive. Indian J. Med. Res., **79**, 131 (1984).

284. SINNIAH, D., and G. BASKARAN: Margosa Oil Poisoning as a cause of Reye's Syndrome. The Lancet I, 487 (1981).

285. SINNIAH, D., G. BASKARAN, P.N. YOEH, and A. RETNASABAPATHY: Treatment of Experimentally induced Margosa Oil Poisoning in Mice. Int. Res. Commun. System, **9**, 114 (1981).

286. SINNIAH, D., P.H. SCWARTZ, R.A. MITCHELL, and E.L. ARCINUE: Investigation of an Animal Model of a Reye-like Syndrome caused by Margosa Oil. Pediatric Res., **19**, 1346 (1985).

287. SRIVASTAVA, S.D.: Limonoids from the Seeds of *Melia Azedarach*. J Nat. Prod., **49**, 56 (1986).

288. SRIVASTAVA, S.D., and S.K. SRIVASTAVA: New Constitutents of *Melia composita*. Fitoterapia, **67**(2), 113 (1996).

289. STILL, W.C., M. KHAN, and A. MITRA: Rapid Chromatographic Technique for Preparative Separation with Moderate Resolution. J. Organ. Chem., **43**, 2923 (1978).

290. STOKES, J.B., and R.E. REDFERN: Effect of Sunlight on Azadirachtin's Antifeedant Potency. J. Environ. Sci. and Health, **A 17**, 57 (1982).

291. SUBRAMANIAN, S.S., and A.G.R. NAIR: Melicitrin, a New Myricetin Glycoside from the Flowers of *Melia azadirachta*. Indian J. Chem., **10**, 452 (1972).

292. SUNDARASIVARAO, B., J. NAZMA, and MADHUSUDHANARAO: Antifungal Activity of Gedunin. Current Sci., **46**, 714 (1977).

293. SURESH, G., N.S. NARASIMHAN, and N. PALANI: Structure of Nimonol from Fresh Whole Green Leaves of *Azadirachta indicaI*. Phytochem., **45**(4), 807 (1997).

294. TAYLOR, D.A.H.: Biogenesis, Distribution, and Systematic Significance of Limonoids in the Meliaceae, Cneoraceae and Allied Taxa. In: Chemistry and Chemical Taxonomy of Rutales. Annual Proceedings of Phytochemical Society of Europe No. 22 (PG Waterman and MF Grundon, eds.), p. 353, New York: Academic Press. 1983.

295. TAYLOR, D.A.H.: The Chemistry of the Limonoids from Meliaceae. In: Progress in the Chemistry of Natural Products **45** (W. HERZ, H. GRIESEBACH, G.W. KIRBY, eds), p. 1. New York, NY: Springer-Verlag, 1984.

296. TAYLOR, D.A.H.: The Chemistry of the Limonoids from Meliaceae. Fortschr. Chem. Organ. Naturstoffe, **45**, 1 (1984).

297. TAYLOR, D.A.H.: Azadirachtin: A Study in the Methodology of Structure Determination. Tetrahedron, **43**, 2779 (1987).

298. TELLA, A.: The Effects of *Azadirachta indica* in Acute *Plasmodium berghei* Malaria. Nigerian Med. J., **7**, 258 (1977).

299. THAKUR, R.S., S.D. SINGH, and A. GOSWAMI: *Azadirachta indica* A. JUSS : A Review. Current Res. Medicinal and Aromatic Plants, **3**, 135 (1981).

300. THIEM, J., H. KARL, and J. SCHWENTER: Synthesis of α-Linked 2′-Deoxy-2′-iododisaccharide. Synthesis, 696 (1978).

301. TIDWELL, T.T.: Oxidation of Alcohols by Activated Dimethyl sulfoxide and Related Reactions – An Update. Synthesis, 857 (1990).

302. THOMPSON, E.B., and C.C. ANDERSON: Cardiovascular Effects of *Azadirachta indica* Extract. J. Pharm. Sci., **67**, 1476 (1978).

303. TEWARI, R.K., R. MATHUR, and A.O. PRAKASH: Post-coital Antifertility Effect of Neem Oil in Female Albino Rats. International Res. Commun. System Med. Sci., **14**, 1005 (1986).

304. TURNER, C.J., M.S. TEMPESTA, R.B. TAYLOR, M.G. ZAGORSKI, J.S. TERMINI, D.R. SCHROEDER, and K. NAKANISHI: A NMR Spectroscopic Study of Azadirachtin and its Trimethyl Ether. Tetrahedron, **43**, 2789 (1987).

305. UDEINYA, I.J.: Anti-Malarial Activity of Nigerian Neem Leaves. Trans. R. Soc. Trop. Med. Hyg., **87**, 471 (1993).

306. UEBEL, E.C., J.D. WARTHEN, and M. JACOBSON: Preparative Reversed-phase Liquid Chromatographic Isolation of Azadirachtin from Neem Kernels. J. Liq. Chromatogr., **2**, 875 (1979).

307. UWAIFO, A.O.: The Mutagenicities of Seven Coumarin Derivatives and a Furan Derivative (Nimbolide) Isolated from Three Medicinal Plants. J. Toxicol. Hlth., **13**, 521 (1984).

308. VAN DER NAT, J.M., L.A. 'T HART, W.G. VAN DER SLUIS, and R.P. LABADIE: Two Functionally Different Immunomodulators from an Aqueous Bark Extract of *Azadirachta indica* A. JUSS (Meliaceae). Pharm. Weekblad Sci. Edition, **9**, 224 (1987).

309. VAN DER NAT, J.M., L.A. 'T HART, W.G. VAN DER SLUIS, H. VAN DIJK, A.J.J. VAN DER BERG, K.T.D. DE SILVA, and R.P. LABADIE: Characterisation of Anti-complement Compounds from *Azadirachta indica*. Ethnopharmacol., **19**, 15 (1989).

310. VAN DER NAT, J.M., J.P.A.M. KLERX, H. VAN DIJK, K.T.D. DE SILVA, and R.P. LABADIE: Immunomodulatory Activity of an Aqueous Extract of *Azadirachta indica* Stem Bark. J Ethnopharmacol., **19**, 125 (1987).

311. VAN DER NAT, J.M., W.G. VAN DER SLUIS, L.A. 'T HART, H. VAN DIJK, K.T.D. DE SILVA, and R.P. LABADIE: Activity-guided Isolation and Identification of *Azadirachta indica* Bark Extract Constituents which Specifically Inhibit Chemiluminescence Production by Activated Human Polymorphonuclear Leukocytes. Planta Medica, **57**, 65 (1991).

312. VAN DER NAT, J.M., W.G. VAN DER SLUIS, K.T.D. DE SILVA, and R.P. LABADIE: Ethnopharmacognostical Survey of *Azadirachta indica* A. JUSS (Meliaceae). J. Ethnopharmacol., **35**, 1 (1991).

313. WARTHEN, J.D.: *Azadirachta indica*: A Source of Feeding Inhibitors and Growth Regulators. Agric. Rev. Manuals ARM-NE-4, p.21, U.S.A.: USDA, Beltsville, MD. 1979.

314. WARTHEN, J.D., J.B. STOKES, M. JACOBSON, and M.F. KOZEMPEL: Estimation of Azadirachtin Content in Neem Extract and Formulations. J. Liq. Chromatogr., **7**, 591 (1984).

315. YAMASAKI, R.B., and J.A. KLOCKE: Structure-Bioactivity Relationships of Azadirachtin, a Potent Insect Control Agent. J. Agric. Food Chem., **35**, 467 (1987).

316. YAMASAKI, R.B., J.A. KLOCKE, S.M. LEE, G.A. STONE, and M.V. DARLINGTON: Isolation and Purification of Azadirachtin from Neem (*Azadirachta indica*) Seeds Using Flash Chromatography and High Performance Liquid Chromatography. J. Chromatogr., **356**, 220 (1986).

317. YAMASAKI, R.B., J.A. KLOCKE, S.M. LEE, G.A. STONE, and M.V. DARLINGTON: Isolation and Purification of Azadirachtin from Neem (*Azadirachta indica*) Seeds using Flash Chromatography and HPLC. Chromatogr., **18**, 467 (1986).

318. YAMASAKI, R.B., T.G. RITLAND, M.A. BARNBY, and J.A. KLOCKE: Isolation and Purification of Salannin from Neem Seeds and Its Quantification in Neem and Chinaberry Seeds and Leaves. J. Chromatogr., **447**, 277 (1988).

319. ZANNO, P.R., E. MIURA, K. NAKANISHI, and D.L. ELDER: Structure of the Insect Phagorepellent Azadirachtin. Applications of PRFT/CWD Carbon-13 Nuclear Magnetic Resonance. J. Amer. Chem. Soc., **97**, 1975 (1975).

320. ZIFFER, H., U. WEISS, and C.R. NARAYANAN: Absolute Stereochemistry of Nimbin. Complex Optical Rotatory Dispersion of Pyronimbic Acid. J. Organ. Chem., **31**, 2691 (1966).

(*Received December 28, 1998*)

Author Index

Page numbers printed in *italics* refer to References

Abe, H. *37–42, 45*
Aburatani, M. *42*
Adam, G. *37–42, 44–46*
Akhila, A. *132, 143*
Allen, E.J. *132*
Altmann, T. *37, 45, 46*
Amberg, W. *43*
Ananich, S.K. *43*
Anazodo, M. *137*
Anderson, C.C. *147*
Anderson, D.M.W. *132*
Anderson, J.C. *132, 133, 140, 146*
Ando, T. *42*
Anna, L.J. *43*
Antonsiewicz, D.M. *46*
Ara, I. *132, 133, 145*
Arcinue, E.L. *147*
Arima, M. *38, 40, 41*
Arimoto, H. *42*
Asakawa, S. *39, 42*
Awasthi, Y.C. *133*
Azpiroz, R. *46*

Baba, J. *39*
Back, T.G. *43*
Bajpai, K.S. *133*
Baker, M.E. *46*
Balandrin, M.F. *133, 140*
Baldwin, J.E. *133*
Banerjee, C.K. *136*
Banerji, R. *133*
Banik, B.K. *133*
Bansal, I. *134*
Banstola, K.P. *142*
Bardhan, J. *143, 146, 147*
Barnby, M.A. *140, 148*
Basak, S.P., *133*
Baskaran, G. *147*
Baskevitch, Z. *143*
Baumann, S. *138*
Bax, A. *133*
Bendall, M.R. *133*
Bennani, Y.L. *43*
Bentley, M.D. *143*
Bhakuni, D.S. *136, 141, 142*
Bhandari, C.R. *145*
Bhargava, K.P. *133*
Bhaskariah, K. *144*
Bhat, S.V. *136*
Bhat, V.S. *144*
Bhide, N.K. *133*
Bhowmick, B.N. *133*
Bilton, J.N. *133*
Blaney, W.M. *133, 140, 141, 146*
Blasko, G. *138*
Böhme, F. *39*
Bokel, M. *134, 138, 139, 141*
Braam, J. *46*
Brandt, W. *45*
Brasca, M.G. *134*
Brosa, C. *45*
Broughton, H.B. *133, 134, 141*
Bruhn, A. 70, *134, 139*
Buchanan, J.G.St.C. 57, *134*
Butterworth, J.H. 103, 111, 116, *134*

Cameron, D.W. *38, 39*
Capdevila, J.M. *45*
Chabra, P.C. *143*
Chadha, R.K. *43*
Chakraborty, D.P. *133*
Champagne, D.E. *134*
Chandrasekharan, V. *133*
Chatterjee, M. *46*
Chaturvedi, R. *141*
Chaudhuri, S.N. *145*

Chavan, S.R.S. *134*
Chawla, A.S. *134*
Chen, S.-H. *134*
Choe, S. *46*
Choi, Y.-H. *37*, *38*, *46*
Choi, Y.-K. *44*
Chopra, P. *143*
Chory, J. *37*, *46*
Choudhary, B.K. *133*
Chua, N.-H. *46*
Clouse, S.D. *37*, *45*, *46*
Coates, W.M. *134*
Coll, J.C. *143*
Connolly, J.D. *134*
Cook, R.K. *46*
Cook Jr., J.K. *36*
Cordell, G.A. *138*
Corey, E.J. *135*
Cornforth, J.W. *135*
Cornforth, R.H. *135*
Corrigan, J.R. *134*
Craig, D. *134*, *141*
Cramer, R. 78, 79, *134*, *138*, *139*
Crispino, G.A. *43*
Croizier, A. *40*
Cutler, H.G. *37*, *45*
Czoppelt, C. *143*

Damishefsky, S.J. *134*
Darlington, M.V. *148*
Datta, P.C. *144*
Davis, W.D. *43*
Dedios, A. *43*
De Groot, A.E. *37*
Delapradilla, R.F. *43*
Denholm, A.A. *137*, *140*
Deshpande, V.Y. *135*
De Silva, K.T.D. *148*
Dess, D.B. *135*
Devakumar, C. *135*, *143*
Dhawan, A.K. *147*
Dilkes, B.P. *46*
D'Incan, E. *135*
Dixit, V.P. *135*
Doddrell, D.M. *133*
Doherty, K. *140*
Dreyer, D.L. *135*
Drosihn, S. *45*
Duddeck, H. *138*
Dutta, P.C. *144*

Easwara, J.P. *132*
Ekanem, O.J. *135*
Ekong, D.E.U. *135*, *136*
Ekundayo, O. *136*
Elder, D.L. *148*
Elich, T. *46*
Ermel, K. *145*
Ezeukwu, G.C. *142*

Facunle, C.O. *135*
Faizi, S. *132*, *133*, *145*, *146*
Fankhauser, C. *46*
Fasina, A.K. *135*
Feldman, K.A. *45*, *46*
Fisher, M.J. *134*
Flippen-Anderson, J.L. *36*
Flood, T.C. *145*
Forster, H. *143*
Fort Jr., R.C. *143*
Frauenglass, E. *141*
Freeman, R. *136*
Friebe, A. *40*
Fry, S.C. *46*
Fuchs, S. *145*, *146*
Fujimoto, Y. *41*, *42*
Fujioka, S. *37–39*, *43*, *44*, *46*
Fujiwara, T. *136*
Fukui, Y. *39*
Fung, S. *42*
Futatsuya, F. *45*

Gagnaire, D. *136*
Gaikwad, B.R. *136*
Gaitande, B.B. *136*
Galbraith, M.N. *44*
Gamoh, K. 21, *38*, *39*, *41*, *42*
Gandhi, M. *136*
Ganeshraj, S.P. *136*, *137*
Garg, H.S. *136*, *141*, *142*
Gasch, A. *46*
Ghatak, U.R. *133*
Ghiasuddin *145*, *146*
Ghosh, S. *133*
Glinsukon, T. *136*, *144*
Goad, L.J. *42*
Godrej, N.B. *45*
Gopalakrishnan, G. *136*, *138*
Goswami, A. *147*
Goswami, B.K. *135*
Gottlieb, H.E. *143*

Govindachari, T.R. *136–138*
Gozalea-Sierra, M. *138*
Griesinger, C. *138*
Griffith, W.P. *137*
Griffiths, P.G. *38*, *39*
Grimminger, W. *138*
Grossman, R.B. *137*
Grove, M.D. 2, *36*, *37*
Gupta, G.P. *133*
Gupta, M.B. *133*
Gutzeit, H. *134*, *139*

Hafferl, W. *44*
Hagaman, E.W. *143*
Hahl, R.W. *135*
Hai, T. *44*
Hall, A.F. *46*
Halsall, T.G. *134*
Hamada, K. *45*
Hammouda, F.M. *40*
Handa, K.L. *134*
Harada, A. *38*, *44*
Harris, M. *137*
Hartung, J. *43*
Haung, H.P. *137*
Henderson, R. *137*
Hendlemeier, R. *138*
Hendrie, A. *132*
Higuchi, K. *38*
Himmelreich, U. *41*
Hiyama, T. *137*
Honjo, C. *39*
Horn, D.H.S. *44*
Horvath, R.F. *134*
Huang, L. *43*
Hyun, K.-H. *38*, *40*

Ibiyemi, S.A. *135*
Ikeda, M. *37*, *38*
Ikekawa, N. *37*, *38*, *40–42*, *45*
Ilavazhagan, G. *143*
Inoue, T. *38*, *43*, *44*
Ishibashi, M. *39*
Ishiguro, M. *42*
Isman, M.B. *134*, *137*, *138*
Iwata, M. *137*
Iwu, M.M. *137*
Iyer, K.N. *142*

Jacobson, M. *137*, *148*
Jain, A.K. *146*
Jain, M.K. *139*, *140*
Jain, R.K. *146*, *147*
Jegede-Fagunsin, H.A. *142*
Jeong, K.-S. *43*
Jessop, A. *46*
Jiang, B. *43*
Jogler, J. *134*
Johnson, S. *132*
Jones, I.W. 109, *137*
Jones, P.S. *133*, *137*, *140*
Joshi, V.S. *144*
Junnarkar, A.Y. *146*

Kabaleeswaran, V. *138*
Kain, A.K. *143*, *147*
Kamalakshi, S. *143*
Kaminski, Z. *137*
Kamuro, Y. *38*, *45*
Karl, H. *147*
Kastigir, H.N. *145*
Kato, N. *41*
Kauschmann, A. *46*
Keil, E. *138*
Keller, U. *138*
Kessler, H. *138*
Ketkar, C.M. *138*
Khalid, S.A. *138*
Khan, M. *147*
Khripach, V.A. *37*, *43*
Kigodi, P.G.K. *138*
Kim, S.-J. *41*
Kim, S.-K. *39*, *40*
Kimura, A. *138*
Kirson, I. *140*
Kitsuwa, T. *37*, *38*, *42*
Klaiber, I. *139*
Klenk, A. *138*, *139*
Klenlen, J.C. *142*
Klerx, J.P.A.M. *148*
Klingele, M.P.H. *138*
Klocke, J.A. *133*, *140*, *148*
Koarai, T. *41*
Koba, S. *39*, *40*
Kobayashi, K. *45*
Kobayashi, K.K. *137*
Koga, Y. *138*
Koguchi, M. *45*
Kolb, H.C. *138*
Kolbe, A. *44*
Koncz, C. *45*, *46*

Koncz-Kálman, Z. *46*
Kondo, H. *42*
Koot, W.-J. *138*
Kornule, P.M. *142*
Kosaka, Y. *40*
Koseki, N. *42*
Koshioka, M. *39*
Kotsuki, T. *142*
Koul, O. *134*, *138*
Koul, R. *137*
Kovganko, N.V. *43*
Kozempel, M.F. *148*
Kraus, W. 69, 70, 78, 79, 112, 116, *134*, *138*, *139*, *141*
Krauss, G. *44*
Kubo, I. 111, *139*
Kuhnt, C. *37*
Kulatilleke, C.P. *141*
Kulkarni, M.M. *144*
Kumar, Ch.S.S.R. *139*
Kumar, M. *134*
Kuriyama, H. *46*
Kyokawa, Y. *39*

Labadie, R.P. *148*
Lachaise, F. *44*
Lafont, R. *44*
Lakhvich, F.A. *43*
Lakshmi, S.U. *139*
Lal, R. *136*
Langford, M. *46*
Lavie, D. 57, *139*, *140*
Lavy, E.C. *140*
Lee, S.M. *133*, *140*, *148*
Leischner, K. *144*
Leuschner, K. *144*
Lewis, H.A. *133*
Ley, S.V. 112, 118, 124, 126, *132–134*, *137*, *138*, *140*, *141*, *146*
Li, J. *37*, *46*
Lidert, Z. *133*, *140*
Little, C.H.A. *40*
LoCascio, J.C. *46*
Loosli, H.R. *138*
Lovell, H. *137*, *141*
Lovell, P.J. *140*, *141*
Lucke, J. *145*, *146*
Luczynski, A. *137*
Lusby, W.R. *43*
Luscombe, D.K. *141*
Lyons, C.W. *141*

Maas, C. *46*
Madhusudanan, K.P. *141*, *142*
Madhusudhanarao, *147*
Madyastha, K.M. *143*
Mahmood, T. *145*, *146*
Maiti, D.C. *141*
Majumdar, P.L. *141*
Makinde, J.M. *142*
Mancuso, A.J. *141*
Mandava, N.B. *36*, *37*, *43*
Marino, J.P. *43*
Marquardt, V. *37*, *42*
Marthandamurthy, M. *144*
Martin, J.C. *135*
Marumo, S. *37*, *38*, *41*, *42*
Massiot, G. *42*, *140*
Mathur, J. *46*
Mathur, R. *143*, *147*
Matsumoto, A. *139*
Matsumoto, T. *139*
Matsuoka, T. *41*
Mayilvaganan, T. *136*
McCrindle, R. *134*, *137*
McMorris, T.C. 24, *43*, *46*
Mehta, D.J. *133*
Meinwald, J. *141*
Melera, A. *137*
Mendulkar, K.N. *135*
Merzweiler, K. *45*
Meudt, W.J. *43*, *44*
Middleton, E.J. *44*
Misra, G. *133*
Mistry, K.B. *45*
Mitchell, R.A. *147*
Mitra, A. *147*
Mitra, C.R. *133*, *141*, *146*
Miura, E. *148*
Mohato, S.B. *143*
Molnár, G. *46*
Mordue, J. *132*
Morgan, E.D. 103, 111, 116, *132–134*, *137*, *140*, *141*
Mori, K. *40*, *42*, *43*, *45*, 126, *141*
Morisaki, M. *42*
Morishita, T. *37*, *38*, *40–42*, *45*
Morita, M. *40*
Morris, G.A. *133*, *136*
Motegi, C. *38*

Mouri, T. *42*
Mubarak, A.M. *141*
Muckensturm, B. *142*
Mukherjee, S. *133*, *141*
Mukherjee, S.K. *135*
Mukherji, A. *144*
Munakata, K. *37*
Munro, A.C. *132*
Murofushi, N. *39*, *43*, *44*
Murty, K.S. *141*
Murty, L.B.G. *141*

Nacpil, Z.D. *143*
Nagasampagi, B.A. *144*
Nagi, G. *139*
Nagpal, P. *46*
Nagy, F. *46*
Nair, A.G.R. *147*
Nakagawa, S. *38*
Nakamura, K. *40*, *42*
Nakamura, S. *39*
Nakanishi, K. *41*, 111, 116, *141*, *145*, *148*
Nakayama, M. *38–41*, *46*
Narasimhan, N.S. *142*, *147*
Narayanan, C.R. *142*, *149*
Narushima, M. *39*
Nath, B. *142*
National Research Council *142*
Natividad, G.M. *143*
Natsume, M. *39*
Nayak, B.R. *142*
Nazma, J. *147*
Neff, M. *46*
Németh, K. *46*
Nieh, M.T. *145*
Nigam, S.K. *133*
Nishide, K. *137*
Nishikawa, N. *39*
Nishiyama, F. *41*
Nishizawa, N. *46*
Noguchi, T. *37*, *46*
Nomura, T. *38*, *39*, *46*, *136*
Nozaki, H. *144*
Nukina, M. *38*
Nuzillard, J.-M. *140*

Obaseki, O. *142*
Obidoa, O. *137*
Obih, P.O. *142*
Ochi, M. *142*
Ogihara, Y. *136*
Ogino, Y. *40*, *42–44*
Ohrai, H. *137*
Okada, K. *40*, *43*
Okamoto, N. *41*
Okogun, J.I. *135*
Okpako, D.T. *142*
Okpanyi, S.N. *142*
Okuda, R. *41*
Olagbemi, E.O. *135*, *136*
Olsen, J.L. *140*
Omote, K. *39*
Omote, N. *41*
Otera, J. *144*
Overton, K.H. *134*, *137*

Pachapurkar, R.V. *142*
Padhya, M.A. *143*
Palani, N. *147*
Pandey, G.N. *141*
Pant, N. *142*
Paramasivam, T.V. *143*
Parashuraman, A.R. *143*
Park, J.-D. *38*, *40*
Park, K.-H. *38*, *40*, *41*
Parmar, B.S. *143*
Patil, P.A. 24, *43*
Pattabiraman, T.N. *139*, *142*
Pegg, D.T. *133*
Pepper, A. *46*
Peracaula, R. *45*
Percy, G.R. *134*
Perichon, J. *135*
Petzold, U. *45*
Pezzuto, J.M. *138*
Pfileger, D. *142*
Pharis, R.P. *40*
Pillai, N.R. *142*, *143*
Piyachaturawat, P. *136*
Plattner, R.D. *37*
Podder, G. *143*
Pohnl, H.D. *138*, *139*
Polisensky, D.H. *46*
Polonsky, J. *134*, *143*
Poole, D. *46*
Popjak, G. *135*
Porzel, A. *37*, *38*, *42*, *44*, *45*
Pradhan, S.K. *142*
Prakash, A.O. *143*, *147*

Prescott, M.C. *42*
Purugganan, M.M. *46*

Ragasa, C.Y. *143*
Raghunathan, R. *136*
Rajab, M.S. *143*
Rajan, S.S. *136*, *138*
Ramanujalu, G. *143*
Ramaprasad, G. *144*
Ramesh, K. *143*
Ramji, N. *143*
Ramraj, V.M. *45*
Rani, K. *132*, *143*
Rao, A.R. *143*
Rao, B.S. *143*
Rao, D.K. *141*
Rao, D.N. *141*
Rao, D.V.K. *143*
Rao, J.M. *143*
Rao, K.N. *143*
Rao, P.U. *144*
Reddi, G.S. *146*
Rédei, G.P. *46*
Redfern, R.E. *147*
Reeb, S. *134*
Reed, J. *46*
Reid, J.B. *41*, *46*
Rembold, H. *143*
Retnasabapathy, A. *147*
Riar, S.S. *143*, *146*, *147*
Rideout, J.A. *143*
Ritland, T.G. *148*
Robert, P.C. *142*
Robinson, N.G. *140*
Rochanakij, S. *144*
Rodriguez-Santamarta, C. *45*
Rojanapo, W. *144*
Rojatkar, S.R. *144*
Rowedder, W.K. *36*
Rukmini, C. *144*
Rzepa, H.S. *133*

Sadio, H. *139*
Sadre, N.L. *135*
Saimoto, H. *38*, *43*, *44*
Saito, H. *37*
Saito, T. *43*
Sakakibara, M. *40*
Sakurai, A. *37–39*, *41*, *43*, *44*, *46*
Sandhya, G. *136*, *137*
Sankaram, A.V.B. *144*
Sankaranarayanan, A. *136*
Santafianos, D. *132*, *137*, *140*, *141*
Santhakumari, G. *142–144*
Sanyal, M. *144*
Sarkar, M.S. *144*
Sassa, T. *37*, *38*
Sasse, J. *45*
Sasse, J.M. *37–39*
Sato, T. *144*
Satoh, H. *45*
Sawaikar, D.D. *144*
Sawamoto, H. *42*
Sawant, B.M. *142*
Sawitzki, G. *139*
Schell, J. *46*
Schimuzu, M. *136*
Schmidt, G. *135*
Schmidt, J. *37–42*, *44*
Schnabl, H. *40*
Schneider, B. *37*, *44*
Schneider, B.H. *145*
Schneider, J.A. *41*
Schroeder, D.R. *145*, *148*
Schweizer, M.P. *140*
Schwenter, J. *147*
Schwinger, M. *138*, *139*
Scudder, G.G.E. *134*
Scwartz, P.H. *147*
Seebach, D. *145*
Sengupta, P. *145*
Sengupta, S.K. *145*
Seshadri, C. *142*
Seto, H. *46*
Shah, V.R. *142*
Shanta, M. *143*
Sharma, H.C. *144*
Sharma, P.L. *136*
Sharpless, K.B. *43*, *145*
Sheppard, R.N. *132*, *133*, *140*
Sheth, V.K. *136*
Shibasaki, H. *144*
Shibasaki, M. 126, *144*
Shibata, T. *43*
Shimada, K. *38–40*
Shoolery, J.N. *139*
Shpan-Gabrielith, S.R. *139*
Shukla, R. *145*
Sibille, S. *135*
Siddall, J.B. *42*, *44*

Siddiqui, B.S. *132, 133, 145, 146*
Siddiqui, S. *132, 133, 145, 146*
Simmonds, M.S.J. *133, 140, 141, 146*
Simonis, M.T. *142*
Sinden, R.E. *137*
Singh, B. *143*
Singh, K. *143*
Singh, K.V. *146*
Singh, N. *45*
Singh, P.P. *146*
Singh, R. *143*
Singh, S. *145*
Singh, S.D. *147*
Sinha, K.C. *146, 147*
Sinha, R. *135*
Sinniah, D. *138, 147*
Sitaramaiah, S. *144*
Slawin, A.M.Z. *133, 134, 140, 141*
Smith, P.M. *39*
Smith, S.C. *133, 140, 141*
Sneen, R.A. *135*
Soca, L. *45*
Sodeoka, M. *144*
Soellner, R. *139*
Somjaree, R. *136, 144*
Somovilla, A.A. *134, 141*
Sonnenbichler, J. *143*
Spaulding, D.W. *43*
Spencer, G.F. *36*
Spengler, B. *40–42*
Spiff, A.I. *136*
Spuck, K. *39*
Srinivas, M. *139, 142*
Srivastav, M. *132*
Srivastava, A.N. *133*
Srivastava, H.C. *141*
Srivastava, S.D. *147*
Srivastava, S.K. *147*
Steffens, G.L. *36*
Steffens, R. *139*
Stephen, J. *144*
Still, W.C. *147*
Stokes, J.B. *147, 148*
Stoldt, M. *45*
Stone, G.A. *148*
Strack, D. *44*
Stuchbury, T. *132*
Subramanian, S.S. *147*
Subramanyam, M. *144*
Sugantla, D. *142*
Suggs, J.W. *135*
Sugishita, E. *136*
Sukumar, S. *143*
Sultana, N. *144*
Sundarasivarao, B. *147*
Suresh, G. *136, 147*
Suwanno, S. *144*
Suzuki, H. *39, 43, 44*
Suzuki, Y. *40, 41*
Swami, B.N. *45*
Swern, D. *141*
Syono, K. *38*
Szekeres, M. *45, 46*

Tada, M. *143*
Tada, T. *142*
Taha, S.A. *141*
Takahashi, N. *36, 38–44*
Takahashi, T. *46*
Takahito, N. *41*
Takatsuto, S. *37–46*
Takeda, T. *136*
Takematsu, T. *39, 41, 45*
Takeuchi, Y. *39, 41, 45, 46*
Tank, R. *135*
Taylor, D.A.H. 112, *147*
Taylor, D.R. *141*
Taylor, P.E. *39*
Taylor, R.B. *148*
Taylor, S.L. *37*
Teijima, I. *41*
Tella, A. *147*
Tempesta, M.S. *148*
Tenjai, C. *144*
Terahata, T. *39*
Termini, J.S. *148*
Terricabras, E. *45*
Tewari, R.K. *143, 147*
Thakur, R.S. *147*
'T Hart, L.A. *148*
Thebtaranonth, Y. *136, 138, 144*
Thiem, J. *147*
Thomas, P. *143, 146, 147*
Thompson, E.B. *147*
Thompson, M.J. *43*
Thomson, J.A. *44*
Thornton, M.D. *141*
Tian, W. *43*
Tidwell, T.T. *147*
Tikadar, S. *144*

Tiwary, R.S. *147*
Tokoroyama, T. *142*
Tomita, Y. *136*
Toogood, P.L. *134*, *140*, *141*, *146*
Towers, G.H.N. *134*
Turner, C.J. *148*
Turner, D.W. *137*

Uchiyama, M. *37*, *38*, *40–42*
Udeinya, I.J. *148*
Uebel, E.C. *148*
Umbreit, M.A. *145*
Uwaifo, A.O. *148*

Van der Berg, A.J.J. *148*
Van der Nat, J.M. *148*
Van der Sluis, W.G. *148*
Van Dijk, H. *148*
Venkatakrishnan, K. *143*
Vincenden, M. *136*
Vitard, V. *46*
Voelter, W. *145*, *146*
Vogler, B. *139*
Voigt, B. *37*, *40–42*, *44–46*
Volz, A. *40*
Vyas, B.A. *136*
Vyas, B.N. *45*

Wachendorff, U. *139*
Wada, K. *42*
Wadia, M.S. *142*
Wagner, C. *45*
Waheed, T.N. *146*
Wakisaki, T. *41*
Warthen, J.D. *148*
Warthen Jr., J.D. *36*
Watabe, Y. *42*
Watanabe, H. 126, *141*
Watanabe, T. 22, 23, *38*, *43*, *46*
Watanabe, S. *42*, *144*
Weiss, U. *149*
Wendisch, D. *139*
Wenkert, E. *143*
Whitcombe, G.P. *137*
White, A.D. *137*
Williams, D.J. *133*, *134*, *140*, *141*
Willmitzer, L. *46*
Winter, J. *44*
Wood, A. *133*, *137*, *140*, *141*
Worley, J.F. *36*
Wu, Y. *46*

Xu, W. *46*

Yamaguchi, I. *40–42*
Yamasaki, R.B. *140*, *148*
Yamashita, F. *138*
Yanagisawa, T. *43*, *44*
Yasuta, E. *39*
Yazawa, N. *45*
Ying, B. *42*
Yoeh, P.N. *147*
Yokota, T. *36–46*
Yoshida, I. *138*
Yoshida, S. *37*
Yoshihara, K. *41*
Yoshino, M. *138*
Yukkundi, S. *139*

Zagorski, M.G. *148*
Zaki, A.K. *40*
Zamora, I. *45*
Zanno, P.R. *148*
Zarbock, J. *138*
Zeigan, D. *42*
Zhabinskii, V.N. *37*, *43*
Zhao, Y.-J. *45*
Zhou, W. *43*
Zhou, W.-S. *43*
Zhou-Halwart, Y. *139*
Ziffer, H. *149*

Subject Index

[2-^{14}C]-Acetate 86, 87
Acetic acid 118, 120, 121
Acetic anhydride 116, 120
Acetobromohydrin 52
11-Acetoxyazadirachtin 113, 117
23-Acetoxy-$\Delta^{20,22}$-azadirachtin 115
3-Acetoxyazadirachtol 114
15-Acetoxy-7-deacetoxydihydroazadirone 66
11-Acetoxy-22,23-dihydroazadirachtin 113, 115, 122
23α / β-Acetoxy-22,23-dihydroazadirachtin 119
23-α / β-Acetoxy-22,23-dihydroazadirachtol 114, 120
23α-Acetoxy-22,23-dihydro-7-keto-11-methoxyazadirachtin 115, 120
23α-Acetoxy-22,23-dihydro-11-methoxyazadirachtin 115
3α-Acetoxy-1α-hydroxyazadirachtol 98, 99
7α-Acetoxymeliaca-14,20,22-trien-3-one 63
11-O-Acetylazadirachtin 105
3-Acetylazadirachtol 120
21-Acetylisonimolicinolide 61
7-Acetylneotrichilenone 62, 65
6-*O*-Acetylnimbandiol 88, 89
3-Acetylnimbidic acid-1,12-lactone 85
6-Acetylnimocinol 64
3-Acetyl-1-tigloylazadirachtinol 124
20-Acetyl-3-tigloylazadirachtol 113, 116, 117
1-Acetyl-7-tigloylnimbidinin 74, 77
1-Acetyl-7-tigloylvilasinin 74, 76
3-Acetyl-7-tigloylvilasinin lactone 79
Aldobiouronic acid 130
Aldotriouronic acids 130
Allylic acetate 120
Alnus glutinosa 9, 21
m-Aminophenylboronic acid 21
Amoebicidal activity 49
Angiospermae 8
Angiosperms 2
Antiallergic activity 49
Antiarthritic activity 62, 109
Antibacterial activity 86
Anticancer activity 86
Anticomplement activity 49
Antidermatic activity 49
Antidiabetic activity 49
Antieczemic activity 49
Antifeedant activity 49, 74, 84, 110, 111
Antifungal activity 49, 71
Antifuruncular activity 49
Antigametogenesist activity 110
Antigingivitic activity 49
Antiinflammatory activity 49, 62
Antimalarial activity 49, 71, 86, 109
Antimicrobial activity 49
Antimiotitic activity 49
Antiperiodontitic activity 49
Antipyretic activity 49
Antipyrrhoeic activity 49
Antiscabic activity 49
Antiseborrhoeic activity 49
Antitubercular activity 49
Anti-ulcer activity 62
Antiviral activity 49
Apiaceae 8, 9
Apium graveolens 9
Apocynaceae 8, 9
Apoeuphol 56
Apo-protolimonoids 50, 51, 56–59
Apotirucallol 56
Arabidopsis sp. 35, 36
Arabidopsis thaliana 9, 34–36
L-Arabinose 130
Arachidic acid 131
Arecaceae 8, 10

Asteraceae 8, 10, 11
Atmospheric pressure chemical ionization mass spectrometry 21
Auxins 13
Azadirachnol 56, 57
Azadirachta indica 48, 51, 74, 81
Azadirachtanin 75, 80
Azadirachtin 88, 98, 100, 103–105, 109–116, 118–121, 124–126
Azadirachtin A 104
Azadirachtin B 97, 98, 104
Azadirachtin D 100, 101
Azadirachtin E 103, 106
Azadirachtin F 101
Azadirachtin G 98, 99
Azadirachtin H 103, 108
Azadirachtin I 100, 101
Azadirachtin K 104, 108
Azadirachtinins 50, 100, 102, 114, 118, 124
Azadirachtins 50, 51, 88, 93, 100, 105–109
Azadirachtol 56, 57, 98, 114, 118, 120
Azadirachtolide 74, 80
Azadirachtols 50, 51, 97–100
Azadiradione 62, 63, 65
Azadirinin 71, 72
Azadirol 56, 57
Azadirone 50, 62–64, 69, 72
Azadirones 50, 51, 64–68

Bactericidal activity 49
Baeyer-Villiger oxidation 24, 27
Baeyer-Villiger reaction 27
Banksia grandis 9
Bean second internode assay 13
Behenic acid 131
4α-Benzoylnimbandiol 88, 90
7-Benzoylnimbocinol 63, 67
Beta vulgaris 9
Betulaceae 8, 9
Biological activity 2, 22, 33, 49, 50, 85, 98, 103, 109, 111, 118
1,4-Bis(9-O-dihydroquinidinyl)-phthalazine 22
Boraginaceae 8, 9
Boronic acid 21
Brassica campestris var. *pekinensis* 9
Brassica napus 2, 9
Brassicaceae 8, 9, 11
Brassinolide 2, 12, 13, 15, 18, 21–25, 27, 29, 31, 35, 36
Brassinosteroid biosynthesis 26–28, 35
Brassinosteroid metabolites 21
Brassinosteroids 2–18, 20–22, 25, 27, 29, 33–36
Butyrospermol 54–56, 58
[^{3}H]-Butyrospermol 87

Campestanol 25, 34
Campesterol 22, 25, 34
[^{13}C]-Campesterol 25
[^{14}C]-Campesterol 25
Cancer 109
Cannabaceae 8, 9
Cannabis sativa 9
Cardioctonic activity 49
Cardiovascular disorders 109
β-Carotene 130
Caryophyllaceae 8, 10
Cassia tora 9
Castanea crenata 9
Castasterone 3, 12, 15, 20–22, 24, 25, 27, 29
(+)-Catechin 130
Catharanthus roseus 9, 25, 27, 29, 35
Cathasterone 34
[^{2}H]-Cathasterone 27
Chemical ionization mass spectrometry 20
Chenopodiaceae 8, 9
Chloroform 13
Chlorophyceae 8
Chlorophyta 8
5α-Cholestane 20
Cholesterol 127
Cholic acids 22
Cinnamic acid 74
1-Cinnamoylvilasinin lactone 78
Cistaceae 8, 9
Cistus hirsutus 9
Citrus sinensis 9
Citrus unshiu 9
Convolvulaceae 8, 10
Crinosterol 22
Cryptomeria japonica 9
Cucumis sativus 9
Cucurbitaceae 8, 9
Cupressaceae 8, 9
Cupressus arizonica 9

1-Cyanoisoindole-2-*m*-phenylboronic acid 21
Cyclic tetrasulfides 131
Cyclic trisulfides 131
Cycloeucalenol 127
Cytochrome P450 31, 35

Dansylaminophenylboronic acid 21
Daucus carota ssp. sativus 9
3-Deacetylazadirachtin 103, 105, 106, 118
Deacetylazadirachtinol 97, 104, 111
7-Deacetylazadirone 62, 64
7-Deacetyl-7-benzoylgedunin 70, 72
7-Deacetyl-7-benzoylnimbinin 63, 68
3-Deacetyl-3-cinnamoylazadirachtin 103, 106
3-Deacetyl-11-desoxyazadirachtin 104, 108
Deacetyl-dihydronimbic acid 86
Deacetyl-2,3-dihydronimbic acid 81, 83
7-Deacetylgedunin 70, 71
6-Deacetylisonimbinolide 82, 84
6-Deacetylnimbanal 83
6-Deacetylnimbin 82
6-Deacetylnimbinal 81, 82
6-Deacetylnimbinene 87
6-Deacetylnimbinolide 82, 84
7-Deacetylnimocinol 64
3-Deacetylsalannin 95
Decalin 121, 122
20,22-Dehydroazadirachnol 56, 57
22-Dehydrocampesterol 22
3-Dehydro-6-deoxoteasterone 7, 20, 27
3-Dehydro-24-epiteasterone 29
2′,3′-Dehydrosalannol 93, 96
3-Dehydroteasterone 6, 20
11-Demethoxycarbonylazadirachtin 103, 108
11-Demethoxycarbonyl-11-oxomelia-carpin 101
Demethyl nimbionol 127, 128
Deoxobrassinosteroids 20
6-Deoxobrassinosteroids 18
6-Deoxocastasterone 3, 12, 27
6-Deoxodolichosterone 4
6-Deoxo-24-epicastasterone 6
6-Deoxo-28-homodolichosterone 5
6-Deoxo-25-methyldolichosterone 6
28-Deoxonimbolide 85–87
6-Deoxo-28-norcastasterone 6
6-Deoxoteasterone 7, 27
6-Deoxotyphasterol 7, 27
Deoxyazadirachtolide 74, 75, 80
2-Deoxybrassinolide 6, 19
2-Deoxybrassinosteroids 19
14,15-Deoxy-gedunin derivatives 71
2-Deoxy-25-methyldolichosterone 5
3-Desacetylazadirachtin 114
3-Desacetyl-1-detigloyl-7-keto-11-methoxyazadirachtin 115, 121
7-Desacetylgedunin 71
Desacetylnimbin 81
6-Desacetylnimbinene 89
3-Desacetylsalannin 93
13,14-Desepoxyazadirachtin-A 105, 109
1α-Destigloyl-1α-benzoylazadirachtin 104, 108
Destigloylsalannin diacetate 95
Detigloylazadirachtin 114, 120
1-Detigloylazadirachtin 103, 106
1-Detigloyl-22,23-dihydroazadirachtin 105, 114
1-Detigloyl-1-isobutyroylazadirachtin 103, 107
1-Detigloyl-1-isocaproyl-3-deacetyl-3-epoxymethacroylazadirachtin 103, 107
1-Detigloyl-1-isovaleroylazadirachtin 103, 107
7-Detigloyl-7-methacroyl-11-deacetylnimbolidin A 92
7-Detigloyl-7-methacroyl-15-deacetylnimbolidin B 93
Detigloylsalannol 95
7-Detigloyl-7-senecioyl-11-deacetylnimbolidin A 92
7-Detigloyl-7-senecioyl-15-deacetylnimbolidin B 93
Diabetes 109
11,20-Diacetoxyazadirachtin 113
3,20-Diacetoxyazadirachtol 114
11,20-Diacetoxy-22,23-dihydroaza-dirachtin 117
1,3-Diacetyl-12α-acetoxyvilasinin 74, 77
Diacetyl azadirachtol 57
Diacetyl azadirol 57
1,3-Diacetyl-7-cinnamoylnimbolinin 90, 91

1,3-Diacetyl-7-cinnamoyl-vilasinin 74, 76
1,3-Diacetyl-11,19-deoxo-19-oxomeliacarpin 100, 101
12,23-Diacetyl nimbocinolide 60
6,23-Diacetyl nimocinolide 60
1,3-Diacetyl-7-tigloyl-12-hydroxyvilasinin 79
1,3-Diacetylvilasinin 74, 75
11,20-Dicarbomethoxyazadirachtin 113, 116, 117
11,20-O,O-Dicarbomethoxy-22,23-dihydroazadirachtin 105
Dicots 2
Dicotyledoneae 8
3,24-Diepibrassinolide 31
2,3-Diepicastasterone 3
3,24-Diepicastasterone 4, 31
2,3-Diepi-25-methyldolichosterone 5
cis-3,5-Diethyl-1,2,4-trithiolane 131
trans-3,5-Diethyl-1,2,4-trithiolane 131
Diglycosides 29
22,23-Dihydroazadirachtin 105, 109, 113, 114, 116–119
Dihydroazadirachtol 98
22,23-Dihydroazadirachtol 114, 120
1,2-Dihydroazadiradione 65
1,2-Dihydroazadirone 63
1,2-Dihydro-4α,6α-dihydroxy-A-homoazadirone 69, 70
22,23-Dihydro-11-methoxyazadirachtin 115, 120, 122
22,23-Dihydro-11,20-methoxyazadirachtin 120
22,23-Dihydro-23β-methoxyazadirachtin 103, 107, 110
1,2-Dihydronimbanal 83
Dihydronimbin 81
Dihydronimbinin 63, 68
Dihydroquinidine 4-chlorobenzoate 24
2′,3′-Dihydrotigloylazadirachtin 103, 106
2′,3′-Dihydrotigloylazadirachtol 98
2′,3′-Dihydrotigloyl-22,23-dihydroazadirachtol 99
4α,6α-Dihydroxy-A-homoazadirone 69, 70
2α,3β-Dihydroxy-B-homo-6a-oxa-5α-pregnane-6,20-dione 31
6,8-Dihydroxy-3-methyl-3,4-dihydroisocoumarin 129
7,8-Dihydroxy-3-methyl-3,4-dihydroisocoumarin 129
2α,3β-Dihydroxy-5α-pregnane-6,20-dione 31
2′,3′-Dihydroxy-2′,3′,22,23-tetrahydroazadirachtin 105
11,20-Dimethoxyazadirachtin 113, 116, 117
Dimethylacetamide 24
N,N-Dimethylacetamide 22
4,14α-Dimethyl-5α-ergosta-8,24(28)-dien-3β-ol 127
Diospyros kaki 9
Di-*n*-propyl disulphide 130
Distylium racemosum 9
Diterpenoids 49
1,7-Ditigloyl-3-acetylvilasinin 74, 80
Diuretic activity 49, 109
Docasene 130
Docosane 130
Dolicholide 4
Dolichos lablab 9
Dolichosterone 4, 15

Ebenaceae 8, 9
Echium plantagineum 9
Enneanortriterpenoids 127
11-Epiazadirachtin H 104, 109
17-Epi-azadiradione 62, 66
24-Epibrassinolide 4, 12, 22, 24, 25, 30–32, 34
2-Epicastasterone 3
3-Epicastasterone 3, 29
24-Epicastasterone 4, 12, 24, 30–32
(−)-Epicatechin 130
ξ-Epi-23-dehydrobrassinolide 16
3-Epi-6-deoxocastasterone 3
2-Epi-2-deoxy-25-methyldolichosterone 6
Epigallocatechin 130
Epilachna varivestis 110
2-Epi-25-methyldolichosterone 5
Epinimbic acid 83
4-Epinimbin 83
Epi-nimbocinol 62
17-Epi-nimbocinol 66
24-Episecasterone 7
24-Epiteasterone 29

24-Epityphasterol 29
Epoxyazadiradione 63, 67
Epoxyazadiradiones 70
1β,2β-Epoxynimbinin 63, 68
Equisetaceae 8, 10
Equisetopsida 8
Equisetum arvense 10, 12
(22*E*,24*R*)-5α-Ergosta-2,22-dien-6-one 24
Ergosterol 22, 24, 25
Eriobotrya japonica 10
Erythronium japonicum 10
Ethanol 120
Ethyl acetate 13, 61, 89
Eucalyptus calophylla 10
Eucalyptus marginata 10
Euphol 54–56, 87
[^{3}H]-Euphol 87

Fabaceae 8–11
Fagaceae 8
Fagopyrum esculentum 10, 20
Fast atom bombardment mass spectrometry 21
Fatty acid 31, 131
Ferroceneboronates 21
Formic acid 118
Fraxinellone 131, 132
L-Fucose 130

Gallic acid 130
(+)-Gallocatechin 130
Gas chromatography 15, 20
Gedunin 70, 71, 109
Gedunins 50, 51, 71–73
Ginkgoaceae 8, 10
Ginkgo biloba 10
Ginkgoopsida 8
23-O-β-D-Glucopyranosyl-25-methyldolichosterone 7, 13
23-O-β-D-Glucopyranosyl-2-*epi*-25-methyldolichosterone 7, 13
D-Glucoronic acid 130
D-Glucosamine 130
D-Glucose 130
Gymnospermae 8
Gymnosperms 2
Gypsophila perfoliata 10

Hamamelidaceae 8, 9
Helianthus annuus 10
Heliothis virescens 67, 110
Hexacosane 130
n-Hexacosanol 130
Hexahydronimbin 81
n-Hexane 13, 14
Homoazadirones 50, 51, 69, 70
28-Homobrassinolide 2
28-Homobrassinosteroids 15
28-Homocastasterone 5
28-Homodolicholide 5
28-Homodolichosterone 5, 15, 16
28-Homoteasterone 6, 19
28-Homotyphasterol 7, 19
Horeau's method 54
Hydrodictyaceae 8, 10
Hydrodictyon reticulatum 10, 12
11-Hydroxyazadirachtin B 104, 108
11-β-Hydroxyazadirachtinin 102
17-β-Hydroxyazadiradione 62, 66
17β / α-Hydroxyazadiradiones 70
6-Hydroxyazadirone 62
6α-Hydroxyazadirone 64
γ-Hydroxybutenolide 58
γ-Hydroxybutenolide apo-protolimonoids 51
Hydroxybutenolides 60
γ-Hydroxybutenolides 59, 93
6α-Hydroxycampestanol 25, 27
23α-Hydroxycampestanol 27
3-Hydroxy-6-deoxobrassinosteroids 20
17α-Hydroxy-14,15-deoxy-17-epi-gedunin 71
20*R*-Hydroxy-3,24-diepibrassinolide 31
25-Hydroxy-3,24-diepibrassinolide 33
20*R*-Hydroxy-3,24-diepicastasterone 31
20-Hydroxyecdysones 31
25-Hydroxy-24-epibrassinolide 29, 30
26-Hydroxy-24-epibrassinolide 30
25-Hydroxy-24-epicastasterone 30
26-Hydroxy-24-epicastasterone 30
12-Hydroxyhavanensin 85
4α-Hydroxy-A-homoisomeldenin 69, 70
22α-Hydroxylated steroids 35
5-Hydroxymethylfurfural 131
17β-Hydroxynimbocinol 67
Hydroxytigloylsalannin 95
Hypoglycaemic activity 49

Icosane 130
Immunomodulatory activity 49

Indole acetic acid 131
Indole pyruvic acid 131
Insect antifeedant activity 109, 118
Insect growth inhibitory activity 67, 90, 98, 111
Insect growth regulating activity 84
Insecticidal activity 49, 90, 98
Iodomethane 116
Isoazadirolide 94, 96
3-Isobutyroylazadirachtol 98, 99
Isofraxidin 129
Isomargosinolide 94, 97
Isomeldenin 62, 65
Isomorgolonone 127
Isonimbinolide 82, 84
Isonimbocinolide 60–62
Isonimbolide 61
Isonimocinolide 60
Isonimolicinolide 61, 70
Isonimolide 61
Isorhamnetin 127
Isosalannolide 95
Isovepaol 103, 107

Kaempferol 127
Kaempferol-3-glucoside 127, 129
7-Ketoazadirachtin 121
7-Ketoazadirachtins 121, 124
7-Keto-11-methoxyazadirachtin 115, 120, 121
24-Keto-26-nimbocinoic acid 53
24-Keto-nimbocinone 53
6-Ketonimocinol 64
20-Keto pregnanes 22
Kovats retention indices 20
Kulactone 54

Lamiaceae 8, 10
Lanosterol 54, 55
Larvicidal activity 49
Lauric acid 31
Leguminosae 31
Lignoceric acid 131
Liliaceae 8, 10, 11
Liliopsida 8
Lilium elegans 10
Lilium longiflorum 10, 13, 21
Limbocidin 74, 78
Limbocinin 74, 78
Limbonin 127, 129
Limocin A 59
Limocin B 59
Limocinin 59
Limocinol 53, 54
Limocinone 53, 54
Limonin 49
Limonoids 49–51, 54, 59–62, 68, 73, 88, 94
Linoleic acid 131
Liquid ammonia 24
Lithium 24
Locusta migratoria 110
Lolium perenne 10
Lychnis viscaria 10
Lycopersicon esculentum 10, 30, 31, 33

Magnoliopsida 8
Mahmoodin 70, 72
Mannose 130
Margocetin 129
Margocilin 128
Margocin 128
Margocinin 128
Margolone 128
Margolonone 128
iso-Margolonone 128
Margosinolide 94, 96
Margosinolone 131
Margosinone 131
Margosolone 128
Margosone 128
Mass spectrometry 15
Meldenin 62, 65
Meldenin diol 62, 64
Melia azadarach 81, 92, 93
Melia azadirachta 48
Meliacarpin derivatives 111
Meliacarpinins 50, 100, 102
Meliacarpins 50, 100, 101
Meliaceae 48, 49, 129
Meliacins 49, 54, 74
Melia composita 74, 80
Meliane-meliacins 57
Melianol 52
Melianol acetate 52
Melianone 52, 54
Melianone bromoacetal 52
Melianone lactone 52
Meliantriol 52, 54–56, 58, 68

Melicitrin 127
Metamorphosis-inhibiting activity 110
1-Methacroyl-3-acetyl-7-cinnamoyl-nimbolinin 91
Methanol 13, 118, 122
11-Methoxyazadirachtin 113, 116–118, 121
1α-Methoxy-1,2-dihydronimbinin 63, 68
6-Methoxymellein 129
Methyl acetate 52
11-O-Methylazadirachtin 105
Methylboronates 20
Methylboronic acid 15
25-Methylcastasterone 8
(24*R*)-24-Methyl-5α-cholestan-3-one 34
(24*R*)-24-Methylcholest-4-en-3-one 34
25-Methyldolichosterone 5, 16
24-Methylenecycloartanol 127
24-Methylene lophenol 127
4α-Methyl-5α-ergosta-8,24(28)-dien-3β-ol 127
4-*O*-(4-*O*-Methyl-α-D-glucopyranosyl uronic acid)-D-galactose 130
Methyl grevillate 131
Methyl nimbiol 127, 128
Methyl nimbionone 127, 128
2-Methyltricosane 130
Mevalonate 86, 87
[2-^{14}C]-Mevalonate 87
[2-^{14}C,(^{4}R)4-$^{3}H_1$]-Mevalonate 87
[2-^{4}C, 4R-$^{3}H_1$]-Mevalonic acid 86
Monocotyledoneae 8
Morgocilin 127
Morgocin 127
Morgocinin 127
Morgolone 127
Morgolonone 127
Morgosolone 127
Morgosone 127
Mutagenic activity 86
Myricetin 127
Myricetin-3′-arabinoside 129
Myricetin glycosides 127
Myristic acid 31, 131
Myrtaceae 8, 10

1-Naphthaleneboronic acid 21
Neeflone 66
Nematicidal activity 49
Nicotiana tabacum 27, 29
Nimbaflavone 127, 131
Nimbanal 82, 83
Nimbandiol 88–90
Nimbic acid 81
Nimbidic acid 74, 85, 86
Nimbidin 49, 74, 109
Nimbidinin 74, 77
Nimbidiol 127, 128
Nimbilicin 127, 128
Nimbilin 90, 91, 93
Nimbin 49, 81, 82, 85, 88, 111
Nimbinal 82, 83
Nimbinene 87, 89
Nimbinenes 50, 87, 89, 90
Nimbinic acid 81
Nimbinin 49, 63, 67, 70
Nimbinol 82–84
Nimbinolide 85
Nimbinone 127, 128
Nimbins 50, 51, 73, 82–85, 87
Nimbiol 127, 128
Nimbione 127, 128
Nimbionol 127, 128
Nimbionone 127, 128
Nimbisonol 127, 128
Nimbocetin 131
Nimbochalcin 130, 131
Nimbocidin 127, 128
Nimbocinol 62, 65, 66, 70
Nimbocinolide 60, 62
Nimbocinone 52–54
Nimbolicin 91, 93
Nimbolide 81, 85–87, 109
Nimbolides 50, 86–88
Nimbolidin A 92
Nimbolidin B 91, 92
Nimbolidins 87
Nimbolin A 74, 76, 90
Nimbolin B 90, 91
Nimbolinin B 93
Nimbolinins 50, 51, 88, 90–93
Nimbonolone 127, 128
Nimbonone 128
Nimbosodione 127, 128
Nimbosone 127, 128
Nimocin 62, 64
Nimocinol 56, 60, 62, 64, 74
Nimocinolide 56, 59, 60
α-Nimolactone 127, 129
β-Nimolactone 127, 129

Nimolicinoic acid 63, 67
Nimolicinol 71, 72
Nimolinin 127, 128
Nimolinone 53, 54
Nimosone 127, 128
Nonacosane 130
Nonadecane 130
28-Norbrassinolide 4
28-Norbrassinosteroids 15
28-Norcastasterone 4, 12

Ochinin acetate 85, 87
Octadecane 130
Ohchinolide A 93
Ohchinolide B 85, 91
Ohchinolide-type compounds 87
Oleic acid 131
Ornithopus sativus 10, 31–33
Orthoester Claisen rearrangement 22
Oryza sativa 10, 29
Osmium tetroxide 119
Oxalic acid 131
6-Oxobrassinosteroids 18, 20
6-Oxocampestanol 27
21-Oxo-ohchinolide 92
12-Oxovilasinin 77
Oxysterols 35

Palmitic acid 31, 131
Pentanortriterpenoids 87
Perilla frutescens 10
Periodinane 120
Phalaris canariensis 10
Pharbitis purpurea 10
Phaseolus vulgaris 10, 13, 31
9-Phenanthreneboronic acid 21
Phoenix dactylifera 10
Phytosterol acetates 20
Phytosterol mixtures 22
Phytosterol precursors 25
Phytosterols 12
Picea sitchensis 10
Pinaceae 8, 10
Pinopsida 8
Pinus sylvestris 10
Pinus thunbergii 10
Piscicidal activity 49
Pisum sativum 11, 35, 36
Plasmodium falciparum 109, 110
Poaceae 8, 10, 11
Polygonaceae 8, 10, 11
Potassium permanganate 119
(*Z*)-Prop-1-enylmagnesium bromide 22
Propionic acid 118
Proteaceae 8, 9
Protolimonoids 50–56, 58, 59
Protomeliacins 50
Psophocarpus tetragonolobus 11
Pteridophyta 8
Pyridine 15
Pyridinium chlorochromate 122
Pyridinium dichromate 120
Pyridinium hydrochloride 24
Pyridinium toluene-p-sulfonate 22
Pyronimbic acid 81
1-Pyruvic-22,23-dihydroazadirachtin 114

Quercetin 127
Quercetin-3-galactoside 127, 129

Raphanus sativus 11
Rhamnose 130
Rheum rhabarbarum 11
Rice-lamina inclination test 13
Robinia pseudo-acacia 11
Rosaceae 8, 10
Rutaceae 8, 9

Salannic acid 85
Salannin 74, 85, 93, 94, 111
Salannin diol 95
Salannins 50, 51, 73, 88, 94–97
Salannol 93, 95
Salannol acetate 93, 95
Salannolactame-21 94, 97
Salannolactame-23 94, 97
Salannolide 93, 95, 96
Schistocerca gregaria 110
Scopoletin 129
Secale cereale 11
Secasterone 6, 12, 20
C-Secomeliacins 74
Selected ion monitoring 20
Senecioic acid 74
1-Senecioyl-3-acetylvilasinin 74, 76
1-Senecioyl-3-acetylvilasinin lactol 79
1-Senecioyl-3-acetylvilasinin lactone 78

Sephadex LH-20 13, 14
Sharpless dihydroxylation procedure 24
Silica gel 13, 14
β-Sitosterol 127
Skin infections 109
Sodium borohydride 81, 82
Sodium methoxide 122
Sodium nimbidinate 109
Sodium periodate 119
Solanaceae 8, 10
Solidago altissima 11
Spermatophyta 8
Spermicidal activity 49
Squalene epoxide 54, 55
Stearic acid 131
Stigmasterol 22, 24, 127
Sugiol 127, 128

Taxodiaceae 8, 9
Teasterone 3, 12, 15, 19, 27, 35
Teasterone-3-laurate 3, 13
Teasterone-3-myristate 3, 13, 21
Tetrahydroazadirachtin 110
2′,3′,22,23-Tetrahydroazadirachtin 105, 113, 117, 118
Tetrahydroazadirachtol 57, 98
2′,3′,22,23-Tetrahydroazadirachtol 113
2′,3′,22,23-Tetrahydro-3-tigloylazadirachtol 117, 118
(22*R*,23*R*,24*S*)-2α,3α,22,23-Tetrahydroxy-24-methyl-B-homo-6a-oxa-5α-cholestan-6-one 2
Tetranortriterpenoids 49, 54, 59, 85
Tetratriacontane 130
Theaceae 8, 11
Thea sinensis 11
Tiglic acid 74, 131
7-Tigloyl-12α-acetoxyvilasinin 74, 77
1-Tigloyl-3-acetyl-12α-acetoxyvilasinin 74, 76
1-Tigloyl-3-acetyl-7-cinnamoylnimbolinin 90, 91
1-Tigloyl-3-acetyl-11-hydroxy-11-demethoxycarbonylmeliacarpin 101
1-Tigloyl-3-acetyl-11-hydroxymeliacarpin 100, 101
1-Tigloyl-3-acetyl-11-hydroxy-meliacarpinin 100, 102
1-Tigloyl-3-acetyl-11-methoxyaza-dirachtinin 100, 102, 118
1-Tigloyl-3-acetylvilasinin 74, 76
1-Tigloyl-3-acetylvilasinin lactol 79
3-Tigloylazadirachtol 97, 98, 113, 114, 116, 118, 120
3-Tigloylazadirachtol derivatives 124
Tigloyl chloride 120
3-Tigloyl-13,14-desepoxy-17-hydroxyazadirachtol 98, 99
3-Tigloyl-22,23-dihydroazadirachtol 99
Tirucallol 54–56, 87
[^{3}H]-Tirucallol 87
[^{3}H]-Δ^7-Tirucallol 87
20(*S*)-Tirucallol 54
1,3,20-Triacetoxyazadirachtol 114
1,3,20-Triacetylazadirachtol 120
Triacetylazadirol 57
1,3,7-Triacetylvilasinin 75
Tributyltin hydride 118
Tricyclic diterpenoids 127, 128
Triethylamine 120
Trifluoroperoxy acetic acid 24, 25
Trihydroxybrassinosteroids 12, 19
2α,3β,6β-Trihydroxy-5α-pregnane-20-one 31
7,11,20-Trimethoxyazadirachtin 113, 116, 117, 120
11,20-*bis*-Trimethylsilylazadirachtin 113, 117
1,7,20-*tris*-Trimethylsilylazadirachtin 113
7,11,20-*tris*-Trimethylsilylazadirachtin 113, 117
1,7,20-*tris*-Trimethylsilylazadirachtol 116, 117
Triterpenoids 49
Triticum aestivum 11
Tulipa gesneriana 11
Typhaceae 8, 11
Typha latifolia 11
Typhasterol 3, 12, 15, 19, 27
[^{2}H]-Typhasterol 27

Δ^{22}-Unsaturated sterols 22

Vaginal contraceptive activity 49
Vepaol 103, 107
Vepinin 73, 74, 80
Vepinins 51
Vicia faba 11, 20

Vigna radiata 29
Vilasinin 74, 75, 80
Vilasinin lactols 75
Vilasinin lactones 75
Vilasinin triacetate 74, 75
Vilasinins 50, 51, 72–80

Wheat leaf-unrolling test 13

Xanthophyll 130
Xylose 130

Zea mays 11

MIX
Papier aus verantwortungsvollen Quellen
Paper from responsible sources
FSC® C105338

If you have any concerns about our products,
you can contact us on
ProductSafety@springernature.com

In case Publisher is established outside the EU,
the EU authorized representative is:
Springer Nature Customer Service Center GmbH
Europaplatz 3, 69115 Heidelberg, Germany

Printed by Libri Plureos GmbH
in Hamburg, Germany